THE PENINSULA CAMPAIGN OF 1862

THE PENINSULA

CAMPAIGN OF 1862

A MILITARY ANALYSIS

KEVIN DOUGHERTY

WITH J. MICHAEL MOORE

UNIVERSITY PRESS OF MISSISSIPPI / Jackson

*Publication of this book was made possible
through the support of Dudley J. Hughes.*

www.upress.state.ms.us

The University Press of Mississippi is a member of the Association
of American University Presses.

Charts courtesy of the Department of the Army
Maps courtesy of the Virginia War Museum
Photographs courtesy of Library of Congress, Prints and
Photographs Division

First edition 2005
∞
Library of Congress Cataloging-in-Publication Data

Dougherty, Kevin.
  The Peninsula Campaign : a military analysis / Kevin Dougherty ;
with J. Michael Moore.—1st ed.
    p. cm.
  Includes bibliographical references and index.
  ISBN 1-57806-752-9 (cloth : alk. paper)
  1. Peninsular Campaign, 1862. 2. Strategy—Case studies.
3. Tactics—Case studies. 4. Command of troops—Case studies.
I. Moore, J. Michael (James Michael), 1950– II. Title.

  E473.6.D68   2005
  973.7'32—dc22   2004022378

British Library Cataloging-in-Publication Data available

CONTENTS

PREFACE

On March 17, 1862, Major General George McClellan launched an amphibious movement from Alexandria to Fort Monroe, Virginia. His intention was to turn the Confederate defenses and to advance on Richmond. Upon landing, the Federals enjoyed a four-to-one numerical superiority over the Confederates. In spite of this initial advantage, McClellan quickly ceased offensive operations and instead endeavored to reduce Yorktown by siege.

While McClellan brought up his siege train, General Robert E. Lee, in his role as military adviser to Confederate President Jefferson Davis, set in motion a reconcentration of forces that would allow the Confederates to block McClellan's approach to Richmond. However, Lee's plan required time. The Confederate forces on the Peninsula would have to fight a delaying action to gain this time.

General Joseph Johnston was in command of the Confederate field forces in Virginia, and Major General John Magruder commanded the Army of the Peninsula. On May 3, 1862, Johnston abandoned Yorktown and fought a delaying action back toward Richmond. Major General James Longstreet covered the withdrawal with a sharp rearguard action fought at Williamsburg on May 5.

As Johnston withdrew up the Peninsula, Norfolk was isolated, and the Confederate forces evacuated it on May 9. This left the CSS *Virginia* without a home port and forced the crew to scuttle the ironclad. With the *Virginia* out of the picture, the Federal navy was free to threaten Richmond via the James River. The Federal gunboats got as far as Drewry's Bluff, within eight miles of Richmond, before being repulsed on May 15.

While Major General Stonewall Jackson was keeping Federal forces tied down in the Shenandoah Valley and keeping President Abraham Lincoln worried about the safety of Washington, Confederates and Federals clashed at Seven Pines on May 31 and June 1. The battle highlighted the dangers of misunderstandings that might arise from verbal orders, but its most significant event was that Johnston was wounded and Lee assumed command. With that, the nature of not just the campaign but of the war itself changed.

Lee developed a bold plan to fix McClellan's front while Jackson, having completed his work in the Shenandoah, attacked McClellan's rear. The attack was planned for June 26. Jackson, however, exhausted from his Valley Campaign, would be uncharacteristically slow, and McClellan escaped the trap. The battles of this period are collectively known as the Seven Days. During these battles, McClellan withdrew to Harrison's Landing under the protection of Flag Officer Louis Goldsborough's gunboats. By August 16, the last units of the Army of the Potomac had left Harrison's Landing to meet transportation at Fort Monroe and embark for new fields. The Peninsula Campaign was over.

The Peninsula Campaign's historiography contains many excellent works. Stephen Sears's *To The Gates of Richmond,* Clifford Dowdey's *The Seven Days,* and Douglas Southall Freeman's *R. E. Lee* and *Lee's Lieutenants* represent notable examples. However, none of these books analyzes the campaign in the context of current and enduring military doctrine. This book represents an effort to fill this void. As such, it is intended more for a military than a historical audience. Background history is provided for continuity, but the heart of the book is military analysis.

The Peninsula Campaign lends itself to such a study. It was the largest campaign of the Civil War. It involved army, navy, air, and marine forces and had the potential to be a truly joint operation. The lessons for the student of the military art are many, spanning the strategic, operational, and tactical levels of war. At the strategic level, there is the tension between Lincoln and McClellan and McClellan's inability to grasp Lincoln's grand strategic objective. At the operational level, there is the failure of the Federal command structure to provide the unity of effort needed for joint operations, the superiority of Confederate versus Federal intelligence, McClellan's forfeiture of the advantage initially gained by his amphibious movement, and Lee's difficulty in synchronizing his attacks. At the tactical level, there is the Confederate use of terrain to trade space for time while Lee effected his reconcentration. There is proof again of the superiority of the flank over the frontal attack and the strength of the defense. Even McClellan's withdrawal offers an excellent example of a retrograde action. There are also numerous lessons to be learned for staff officers, especially from the inability of the Confederate staff to help General Lee effectively control the battle.

Moreover, the campaign is resplendent with lessons learned about the personal dimension of war. There is McClellan's overcaution, Lee's audacity, and Jackson's personal exhaustion. All provide valuable insights for today's commanders.

History is most valuable when it is studied for the purpose of learning lessons to be applied in current and future situations. This is the purpose of this volume. It is designed to help today's military leaders examine a historical operation in light of current doctrine and see what may be learned.

THE PENINSULA CAMPAIGN OF 1862

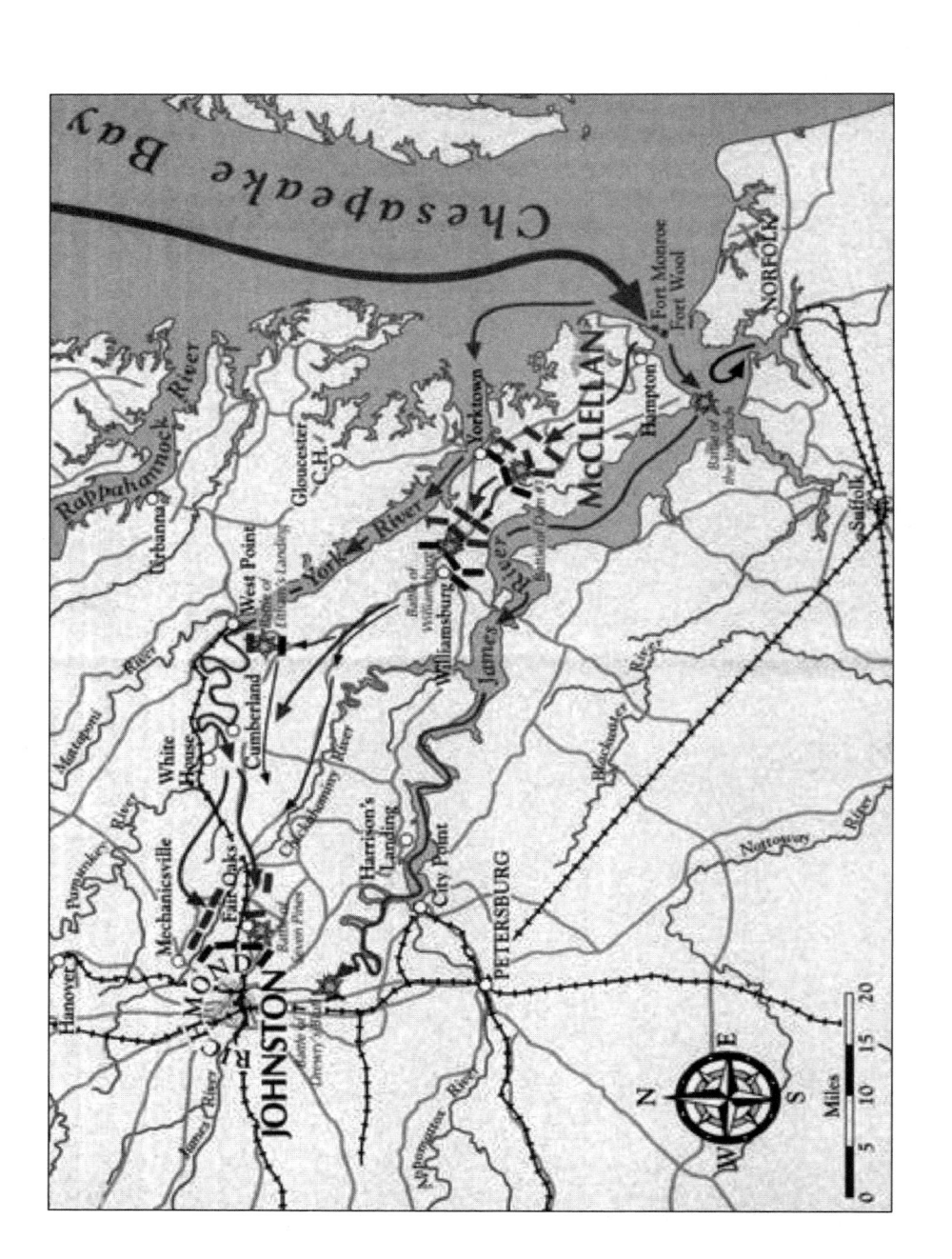

CHAPTER 1

THE STAGE IS SET

Among the many factors that shape a battle's outcome are the individual personalities of the commanders involved and the doctrine they use to prosecute the war. One cannot appreciate the German blitzkrieg of World War II without also appreciating the German doctrine of *Auftragstaktik* or the personality of Heinz Guderian. The American approach to Desert Storm is inexorably intertwined with the AirLand Battle Doctrine and personality of Norman Schwarzkopf. So it is with the Peninsula Campaign, a joint (or multiservice) campaign undertaken in an era with only the most rudimentary understanding of joint operations prosecuted by leaders who represented the full spectrum of experience, capabilities, weaknesses, competencies, visions, and personalities.

Napoleon believed that "The personality of the general is indispensable; he is the head, he is the all, of an army" (Fuller, "Generalship" 30). A discussion of the Peninsula Campaign logically begins with an understanding of these personalities.

THE KEY FEDERALS
WILLIAM BUEL FRANKLIN (1823–1903)

Franklin graduated first in the West Point class of 1843 and received a second lieutenant's commission in the Corps of Engineers. Franklin supervised harbor improvements, earned two brevets in the Mexican War, and taught at West Point. In April 1861, he directed the construction of the new dome on the Capitol. Promoted to brigadier general on May 17, 1861, Franklin commanded a brigade at First Manassas and by October 1861 was a division commander in the Army of the Potomac. He was a confidante of Major General George

McClellan and recommended an amphibious turning movement similar to the one planned by McClellan when queried by President Lincoln in January 1861. Franklin's division served as the strategic reserve during the siege of the Warwick-Yorktown line.

After the capture of Yorktown on May 4, 1862, McClellan belatedly sent Franklin's command up the York River to trap the Confederate Army retreating toward Richmond. The Confederates, however, escaped this flanking maneuver at the battle of Eltham's Landing and continued to withdraw toward Richmond.

Later in May, McClellan received permission to create two new corps and personally to select their commanders. Looking to elevate generals who shared his views, McClellan appointed Franklin to command the Sixth Corps. Franklin held this position during the Seven Days and performed well (Faust 285; Boatner 303–4). Indeed, Clifford Dowdey assesses that at Glendale, "Franklin's cool initiative had been able to bring order out of a confusion of random troop movements and conflicting assignments" (*Seven Days* 275).

LOUIS MALESHERBES GOLDSBOROUGH (1805–1873)

Goldsborough was warranted as a midshipman at age seven and actually entered the service four years later. At the time of the Peninsula Campaign, he had been a sailor almost his entire life. He had made cruises in the Mediterranean and the Pacific, commanded the *Ohio* during the Mexican War, served with Commodore Matthew Perry at Tuxpan in 1847, and been superintendent of the U.S. Naval Academy. In September 1861, he was appointed commander of the North Atlantic Blockading Squadron ("Webster's" 146). Shelby Foote describes Goldsborough as "a big, slack-bodied regular of the type called 'barnacles'" (Foote 228).

Goldsborough would be a critical component for what McClellan had in mind on the Peninsula. McClellan was hoping to achieve the same sort of close army-navy cooperation that Brigadier General Ulysses S. Grant and Captain Andrew Foote enjoyed at Fort Henry in February 1862. Even without any direct or formal command link (Foote had been instructed by Secretary of the Navy Gideon Welles to cooperate with the army without subordinating himself to it), Grant and Foote worked closely together in arranging all aspects of the attack. Jointly, Grant and Foote telegraphed Department of the Missouri commander Major General Henry Halleck with their plan. Indeed, Foote's endorsement won Halleck's approval. On February 6, Foote shelled Fort Henry into submission. Grant's land force arrived thirty minutes after the surrender and took possession of the fort.

Scott Stuckey concludes that "in the absence of unified command or meaningful joint doctrine, the conception and execution of joint operations totally depended on ad hoc actions by the responsible commanders, and therefore upon their personal chemistry and communications. Foote and Grant were very different individuals . . . yet they worked well together. Whatever their differences, they shared a common inclination to attack the enemy, both hating inactivity. They maintained excellent communications without undue worry as to who would get the credit—a quality rare in Civil War commanders" (94).

Unfortunately, the relationship between Goldsborough and McClellan would never produce such "personal chemistry and communications," and the cooperation enjoyed between the army and navy at Fort Henry would never materialize on the Peninsula. The older and portly Goldsborough presented a marked contrast to the young and dashing McClellan. Additionally, Goldsborough's terrible temper (Niven 347) and McClellan's high opinion of himself would present an unwieldy combination. Where the two men were compatible—Goldsborough has been accused of an "overly defensive mentality" (Reed 40), and McClellan was well known for his caution—there was little to suggest action. All this played out against a backdrop of residual tension between the army and the navy over McClellan's earlier failure to heed the navy's advice and launch an operation to seize Norfolk (Glatthaar 70–71).

For his part, Goldsborough had previously participated in such a cooperative environment. He had recently commanded the naval component of Brigadier General Ambrose Burnside's expedition against Roanoke Island off the North Carolina coast. Goldsborough showed great skill and personal bravery in handling this motley assembly of shallow-draft transports of various descriptions and levels of seaworthiness. More importantly, he showed he could cooperate with the army, or at least with Burnside. Indeed, Bern Anderson says the "operation was an excellent example of the coordination that could be achieved by competent commanders" (64).

On January 7, Goldsborough's gunboats moved out ahead of the transports and their thirteen thousand assault troops. Burnside and Goldsborough effectively controlled the naval bombardment to support the land operation (Anderson 63–64). Confederate resistance was ineffective, and the next day Burnside found himself in control of North Carolina's inland sea. Shelby Foote describes the success of the expedition as having produced a "newly bred amphibious beast, like some monster out of mythology—half Army, half Navy: an improbable, unholy combination if ever there was one—[that] might come splashing and roaring ashore at any point" (230). McClellan wanted this next point to be the Virginia Peninsula. Whether he could make this plan a reality would depend in large measure on his ability to cooperate with Goldsborough.

Something, however, seemed to get lost in the translation whenever the two talked. As T. Harry Williams puts it, "After McClellan made his request for naval support, a lot of talking took place about enlisting it, but everybody seems to have misunderstood everybody else" (*Lincoln* 76). The two commanders would fail to agree on a common purpose, with Goldsborough opting to focus on protecting McClellan's base on the Peninsula and the sea lines to it rather than attacking Confederate forces ("Webster's" 146). The Peninsula Campaign was destined to fall far short of being an effective joint army-navy operation.

WINFIELD SCOTT HANCOCK (1824–1886)

Hancock graduated eighteenth in the West Point class of 1844. He fought in the Second Seminole War and the Mexican War. He also served in Kansas during the border war and was a quartermaster officer in California at the beginning of the Civil War. Promoted to the rank of brigadier general, Hancock commanded a brigade in Smith's Division during the Peninsula Campaign.

His brigade was engaged at Lee's Mill, Dam no. 1, and Williamsburg ("Webster's" 337; Boatner 372). Shawn Harris writes that at Williamsburg, Hancock was the Federal army's "one bright spot." William Smith, Hancock's division commander, cited Hancock's exploits in an effort to "earnestly call the attention of the General-in-Chief for just praise." Erasmus Keyes, the Fourth Corps commander, stated, "If Hancock had failed, the enemy would not have retreated." Even McClellan felt that "Hancock was superb" (Harris 48).

SAMUEL PETER HEINTZELMAN (1805–1880)

Heintzelman graduated seventeenth in the West Point class of 1826 and performed garrison duty as an infantry and quarter-master officer. Heintzelman developed a reputation for personal bravery and received two brevets for gallantry in the Mexican War and in the Southwest. On May 17, 1861, he was promoted to brigadier general after thirty-five years of service. Heintzelman later commanded a division at First Manassas and was wounded while rallying his troops. He voted against McClellan's Urbanna Plan at the March 8 council of war. During the Peninsula Campaign, Heintzelman commanded the Third Corps (Faust 356; Boatner 372).

Mark Boatner concludes that Heintzelman was "A man who lacked the essential qualities of leadership as well as one who greatly magnified the difficulties before him; he was personally brave and gallant but without initiative" (392). Bruce Catton agrees that Heintzelman was "brave enough for a dozen men but unfitted for any problem of leadership that extended beyond men he could reach with his own voice" (*Army of the Potomac* 114). By overestimating the strength of the Confederate troops at Yorktown, Heintzelman reinforced McClellan's already cautious personality.

JOSEPH HOOKER (1814–1879)

Hooker, an 1837 graduate of West Point, fought in the Second Seminole War and served as the adjutant at West Point. Lieutenant Hooker received three brevets during the Mexican War but incurred the displeasure of Lieutenant General Winfield Scott for defending Gideon Pillow against charges of insubordination. In 1853, Hooker resigned his commission and moved to California. On May 17, 1861, Hooker was appointed a brigadier general and assigned to the Washington defenses. He commanded a division in the Third Corps during the Peninsula Campaign and prematurely initiated action at Williamsburg (Faust 369; Boatner 409–10). A series of Associated Press releases during the Seven Days were headed "Fighting—Joe Hooker," and newspapers all over the country simply removed the dash and used "Fighting Joe Hooker" as a subhead. The nickname stuck (Boatner 409).

PHILIP KEARNY (1815–1862)

Kearny, the nephew of Colonel Stephen Kearny, received a degree from Columbia University in 1833 and later inherited a million dollars from his grandfather. Despite his wealth, Kearny sought a military career, and by the time of the Peninsula Campaign, he had seen as much fighting as anyone. He had served in the French Army in Algiers in 1840, suppressed the Mexican revolt in New Mexico and southern California in 1846–47, and lost an arm in the capture of Mexico City. He resigned from the U.S. Army and again fought with the French in northern Italy. Unquestionably brave, Kearny, a French officer wrote, "went under fire as on parade, with a smile on his lips." Winfield Scott called him "the bravest man I ever saw" (Catton, *Army of the Potomac* 31).

Appointed a brigadier general on May 17, 1861, Kearny commanded a division during the Peninsula Campaign and performed well at Williamsburg and Seven Pines. Always ready for a fight, at Williamsburg he declared, "I can make men follow me into hell!" (Harris 45). He was promoted to major general on July 4, 1862.

While on the Peninsula, Kearny developed a tradition that continues to today's army—the unit patch. Happening upon a group of officers loitering by the side of the road, Kearny delivered them a stern lecture only to learn that they did not belong to his command. To prevent a repeat of this embarrassing mistake, Kearny instructed all the officers and men of his division to wear a red diamond on their caps or coats. The idea caught on and spread throughout the army (Dougherty, "Scraps" 52).

ERASMUS DARWIN KEYES (1810–1895)

Keyes graduated tenth from West Point in 1832 and performed frontier garrison duty. Keyes later served as Winfield Scott's military secretary and as an artillery instructor at West Point. George McClellan was one of Keyes's students, and he felt that "a pleasanter pupil was never called to the blackboard" (Sears, *McClellan* 7). By 1861, Keyes was colonel of the Eleventh U.S. Infantry and commanded a brigade at First Manassas. Promoted to brigadier general, he received command of the Fourth Corps in March 1862 (Faust 416; Boatner 483–84).

From the outset, Keyes had felt that the Peninsula Campaign could succeed only with naval support. When he encountered Magruder's Warwick River defenses, Keyes's alarmist reports helped influence McClellan to forgo maneuver in favor of siege operations. Dowdey writes that had Keyes been given command "solely for the purpose of drenching McClellan's offensive flare, he could not have been more effective" (*Seven Days* 45).

ABRAHAM LINCOLN (1809–1865)

Lincoln was a self-educated frontier lawyer who served in the Illinois legislature and the U.S. Congress. Although he lost his 1856 bid for a Senate seat against Stephen A. Douglas, Lincoln became a national political figure and debated the westward expansion of slavery. In July 1860, Lincoln received the Republican Party's nomination for president, and the Democratic Party's sectional divisions ensured his election in November 1860. The outcome of this election led to the secession of most of the slaveholding states.

Lincoln, though a militia captain in the Black Hawk War, had little practical military experience. Consequently, McClellan viewed the president condescendingly and resisted his suggestions for immediate military action against the Confederates in northern Virginia. Lincoln reorganized the Army of the Potomac and removed McClellan as general in chief on March 11, 1862. During the Peninsula Campaign, McClellan felt that this move disrupted the Union command structure and limited his power to secure the U.S. Navy's cooperation and reinforcements from other commands.

Lincoln showed great patience with McClellan's proclivity for inaction on the battlefield and overestimation of the Confederate forces, developing a greater appreciation of grand strategy than did the general (Faust 439; Boatner 458). Indeed, Joseph Glatthaar concludes that Lincoln "was a builder, a man possessed of true strategic vision. He focused on the end, a reunited nation, and adapted his ways and means to fulfill it. What McClellan perceived as weakness was in fact flexibility" (70–71).

THADDEUS SOBIESKI CONSTANTINE LOWE (1832–1913)

Lowe was born in Jefferson Mills, New Hampshire, and always had a love of ballooning. With the outbreak of the Civil War, Lowe found himself in competition with John Wise and John La Mountain to convince the Federal army of the military applications of ballooning. Wise would have the distinction of building the first balloon bought for military use in America, and La Mountain would precede Lowe to the Peninsula, but in the end, Lowe persevered and became the father of the U.S. Balloon Corps.

In June 1861, Lowe arrived in Washington with his balloon, the *Enterprise*. Both Secretary of War Simon Cameron and General in Chief Winfield Scott had no time to see Lowe, but the idea of balloon reconnaissance caught the fancy of President Lincoln, and Lowe arranged for a demonstration flight at an altitude of five hundred feet. He rigged a telegraphic system with one terminal in his balloon basket and one at the War Department with an extension to the White House. On June 18, Lowe sent the world's first electric telegraph message from the air. It read:

> To the President of the United States:
> Sir: This point of observation commands an area nearly 50 miles in diameter. The city with its girdle of encampments presents a superb scene. I have pleasure in sending you the first dispatch ever telegraphed from an aerial station and in acknowledging indebtedness for your encouragement for the opportunity of demonstrating the availability of the science of aeronautics in the service of the country.
>
> —T. S. C. Lowe (Richardson 35)

Lincoln wired a reply, and after the flight, the crew towed the balloon through the city streets and anchored it on the White House lawn. The *Enterprise* remained there overnight, and the next day Lincoln gave it a close inspection. He later would personally escort Lowe to see Scott, and the U.S. Balloon Corps was born (Richardson 36).

Lowe eventually would have a total of seven balloons in service (Halsey 81). They were designed to be filled with hydrogen produced in Lowe's field generators, which were wooden tanks carried on wagons and filled with water and iron filings. When the mixture was doused with sulfuric acid, the resulting action would yield hydrogen gas. The gases were cooled in copper pipes passed through the line (Davis, *Civil War* 54). These generators were Lowe's most important invention because they freed the Balloon Corps from reliance on city gas works. Thus, the balloonists could travel with the army wherever it went (Bailey 147). Lowe also had at his disposal what Burke Davis, in spite of rival La Mountain's earlier work with the *Fanny,* refers to as history's first "true aircraft carrier," the USS *George Washington Parke Custis* (*Civil War* 52). The *Custis* was an old coal barge that Lowe had converted to a flattop. He used the barge to haul his tethered balloon up, down, and across the Potomac and James Rivers to get a better view of Confederate positions. By being able both to travel with the army and to operate from a waterborne platform, Lowe's balloons were well suited for joint operations.

Lowe's mission was to gather intelligence. He preferred to observe from a height of three hundred feet (Bailey 151) but on one occasion rose to five thousand feet (Halsey 81). He found that the best time for observation was just before dawn. The air was usually clear at that time, and the observer could readily see the campfires that revealed troop locations and strengths. Changes in the numbers or locations of campfires between two observations would indicate movements (Bailey 151).

If the balloon were operating behind secure battle lines, the observer could gather information with relative leisure. Engineers would make detailed sketches for maps, and observers would make notes that would later be delivered to the commander by messenger. If, however, speed was of the essence, the observer would be linked to the commander by telegraph. In this manner, information about enemy dispositions could be swiftly relayed from air to ground (Bailey 152).

With his notorious hunger for information about enemy troop strengths, McClellan was one of Lowe's most enthusiastic supporters. Thus, when McClellan launched the Peninsula Campaign, Lowe went along with three balloons, his trains, and the *Custis*. In fact, before beginning the campaign, McClellan had Lowe dispatch one balloon to Fort Monroe for the sole purpose of keeping track of the *Virginia* (Richardson 38). McClellan as well as Generals Daniel Sickles, Charles Stone, George Stoneman, Samuel Heintzelman, and Fitz John Porter all made personal ascensions with Lowe or one of his assistants (Richardson 38).

Lowe, however, would not have a balloon aloft the mornings of the attacks at Seven Pines or Mechanicsville and provided no early warning of Confederate attacks at either place (Sears, *Gates* 196). The Federal army lacked an all-source intelligence service, and Lowe's operation was just another independent effort.

However, Lowe's balloons did bring a new dimension to the battle-field, ushering in the era of aerial observation and paving the way for future cooperation between air and ground forces. This potential, how-ever, would not be fully realized on the Peninsula. For all his efforts, Lowe "furnished [McClellan] nothing that brought any reality to the way he was counting the Army of Northern Virginia" (Sears, *Gates* 54). Lowe's balloons were an interesting part of the Federal intelligence effort, but the effort itself remained a failure.

GEORGE BRINTON McCLELLAN (1826–1885)

McClellan graduated second in the West Point class of 1846 and took a commission in the engineers. His classmates included Stonewall Jackson, who graduated seventeenth and who would fight against McClellan some sixteen years later outside Richmond. Like Jackson and many others, McClellan's first combat experience came in the Mexican War. McClellan arrived in Mexico in October 1846 and served under Robert E. Lee and John Magruder, two more of his oppo-nents on the Peninsula, at different times during the war.

However, McClellan's most important experience in Mexico with regard to the Peninsula Campaign would come at Vera Cruz, a thirteen thousand man amphibious turning movement masterminded by General Winfield Scott. On March 9, 1847, McClellan and his company of engineers landed with the first wave of regulars outside of Vera Cruz. The Mexicans failed to oppose the landing, and within four days the Americans had drawn their lines around the fortified city. Scott opted against a frontal assault and instead initiated a formal siege. The original siege train proved to be inadequate, and Scott arranged to borrow heav-ier ordnance from the navy (Sears, *McClellan* 18–19). The shelling began on March 22, a truce was granted on March 26, and on March 27 the Mexican garrison surrendered. It was all relatively quick and painless. It was a lesson McClellan would not forget.

From the Mexican War, McClellan also developed a thorough con-tempt for the civilian management of the war that became embedded in his military thinking (Sears, *McClellan* 25). Part of this conclusion was surely influenced by McClellan's having to serve under the politically appointed and largely incompetent Brigadier General Gideon Pillow and Major General Robert Patterson. An additional factor was what McClellan considered to be President James Polk's politically motivated relief of Scott in 1848.

After the war, McClellan returned to an assignment at West Point, but his most significant peacetime army experience began on April 11, 1855, when he departed as part of the Delafield Commission, a three-man military commission appointed by Secretary of War Jefferson Davis to study the latest military developments in Europe and to observe firsthand the war in Crimea. Upon arrival, the commission was thwarted in its attempts to visit Sevastopol during the siege itself but was able to inspect the abandoned works after the siege was lifted. McClellan spent much of his time studying the logistical aspects of the siege, including the transportation and shipping of the allied forces by sea (Sears, *McClellan* 48). When he returned to the United States, McClellan opened his report with a critical analysis of the siege at Sevastopol. Indeed, no officer knew more about conducting a siege than did McClellan (Sears, *Gates* 38–39). This experience would clearly impact McClellan's strategy and conduct on the Peninsula.

McClellan had also devoted considerable energy to observing how the European professional armies were trained and organized. This experience would greatly enhance McClellan's abilities as a builder of armies and as an administrator. However, McClellan seems to have ignored several basic tactical lessons that were apparent in the Crimea. His report made no mention of the use of the rifled musket and the dramatic impact that this development would have on future warfare. Sears concludes that "as a consequence, neither [McClellan] nor those who read his report were stimulated to rethink tactical doctrine as a result of the Crimean War. Whatever advantage he would have as a military administrator in 1861, he possessed no more tactical insights into the war to come than did any of his fellow (or opposing) generals" (*McClellan* 49).

Perhaps herein lays a clue to McClellan's excellence as an administrator, organizer, and planner of military operations but his serious flaws as an executor. McClellan had a highly systematic type of mind. He liked everything to proceed in accordance with a careful plan (Sears, *Gates* 38). When the situation changed and flexibility was required, as it would be on the Peninsula, McClellan could not adapt. In the words of James Longstreet, "General McClellan's plans were laid according to strict rules of strategy, but he was not quick or forcible in handling his troops" (80).

Furthermore, McClellan's perfectionist tendencies could become paralyzing. Colonel Oliver Spaulding observed that McClellan "was never satisfied with what he had, nor willing to make the best of an imperfect tool. He could always see wherein he might make improvements, given time; and he took time, at the expense of losing his opportunities. . . . His reasoning powers carried him up to contact with the enemy; at that moment, when an independent will entered the problem, he became hesitating" (Hassler, *Commanders* 31). This is exactly how McClellan would behave on the Peninsula.

McClellan's participation in the Delafield Commission was the pinnacle of his peacetime army career, and he seemed content to end on this high note. He resigned from the army on January 15, 1857, and took a position as chief engineer with the Illinois Central Railroad. Within a year he was vice president, and his railroad career culminated as superintendent of the Ohio and Mississippi Railroad. McClellan proved to be an able executive, but he did not really like his new role as a businessman. He made several abortive efforts to find a new line of work or return to a military career.

However, even McClellan's railroad experience would help shape his conduct on the Peninsula. Through the railroad, McClellan met Allan Pinkerton, the private detective under contract with the Illinois Central and other railroads to protect their property. In 1861, McClellan would appoint Pinkerton to be his intelligence chief, and Pinkerton's exaggerated reports of Confederate strengths would have a multiplying effect on McClellan's already cautious disposition.

McClellan's other formative experience during this period was his introduction to Abraham Lincoln. While Joseph Glatthaar argues that McClellan's problem was not with Lincoln in particular but rather with all authority figures (237–42), the fact remains that for whatever reason, McClellan did not like Lincoln. Lincoln had represented the Illinois Central in a number of legal cases, and McClellan was not particularly impressed with the future president's work.

The 1858 Illinois senatorial race pitted Lincoln against Stephen Douglas. McClellan was impressed by Douglas's oratory skills and critical of Lincoln's. McClellan voted for Douglas, the Democrat. The roots of the suspicious and patronizing contempt with which McClellan would later treat President Lincoln can be traced to these earlier experiences in Illinois. Archer Jones writes of McClellan that, "Having known Lincoln from their days in Illinois, the urbane professional soldier apparently had little respect for the country lawyer's ability to discharge his duties as commander in chief" (43).

Thus, at the outbreak of the Civil War, McClellan had had several experiences that would shape his future conduct. He had a gifted intellect, to be sure, and was clearly regarded as an expert on military affairs. He had observed both amphibious turning movements and sieges at Vera Cruz and the Crimea. He had studied military administration and performed similar duties with the railroad. He had also met many of the people he would serve with, under, and against on the Peninsula. Perhaps of all the key players on the Peninsula in 1862, McClellan's behavior is easiest to explain based on his preparatory experiences.

IRWIN McDOWELL (1818–1885)

McDowell, an 1838 graduate of West Point, taught tactics there between 1841 and 1845. McDowell served on the staff of Brigadier General John Ellis Wool during the Mexican War and was brevetted to the rank of captain. The rest of his career was spent as a staff officer in the Adjutant General's Department. On May 14, 1861, McDowell was promoted to brigadier general and given command of the Federal Army of the Potomac. His inexperienced troops were defeated at First Manassas, and McClellan took command of the Army of the Potomac. In March 1862, McDowell was promoted to major general and received command of the First Corps (Faust 459; Boatner 531).

McClellan was always suspicious that McDowell was continually conniving to get his old command back (Sears, *McClellan* 142, 188) and considered McDowell a tool of the abolitionists (Sears, *McClellan* 244). To make matters worse, McDowell had voted against McClellan's Urbanna Plan at the March 8 council of war. Nonetheless, the First Corps was the largest in McClellan's army, and he considered it critical to the success of the Peninsula Campaign. Lincoln, apprehensive of the threat Stonewall Jackson posed to Washington from the Shenandoah Valley, ordered the

First Corps to remain as part of Washington's defenses rather than joining McClellan. McClellan would cite this decision as ruinous to his campaign.

Porter, a cousin of Admiral David Dixon Porter, graduated eighth in the West Point class of 1845 and went on to win brevets of captain and major in the Mexican War. Early in the Civil War, he served as chief of staff to Major General Robert Patterson in the Shenandoah Valley, but McClellan soon called Porter to Washington to help form what would become the Army of the Potomac. From this point on, Porter's destiny would be intertwined with McClellan's.

Porter and McClellan quickly bonded based on both their shared Democratic political leanings and their comparable thoughts on military theory. Politically, Porter felt that "abolitionist traitors" were dividing the North in its war effort. Militarily, he felt that if "McClellan is defeated it will be the fault of the administration, not his own" (Fleek 55).

Because of this like-mindedness, Stephen Sears concludes that "McClellan rated Porter his favorite lieutenant by a wide margin." McClellan demonstrated this sentiment by placing Porter in overall charge of the siege at Yorktown and later by posting him in the critical position north of the Chickahominy River outside Richmond during the Seven Days battles (*Gates* 211).

At the beginning of the Peninsula Campaign, Porter was a division commander in the Third Corps, but on May 18, after McClellan had been authorized to create two additional corps, Porter was elevated to commander of the Fifth Corps. The appointment was critical to McClellan, who needed corps commanders who thought like he did.

Like McClellan, Porter would be accused of being unwilling to fight. He would demonstrate great skill at Mechanicsville, Gaines' Mill, and Malvern Hill, but his critics would note that these were all defensive engagements. Such criticism notwithstanding, by the end of the campaign, Porter would be promoted to major general of volunteers and brigadier general by brevet in the regular army.

However, Porter's good fortune took an abrupt turn for the worse after the Peninsula Campaign. As McClellan found himself rapidly losing favor with Lincoln, many of his troops were withdrawn from his Army of the Potomac and attached to Major General John Pope's Army of Virginia. Porter's men were among this group, and he let his dissatisfaction with the situation be known. Word of this naturally reached Pope.

After his defeat at Second Manassas, Pope preferred court-martial charges against Porter for disloyalty, disobedience, and misconduct in the face of the enemy. In a politically charged atmosphere in which the Radicals saw an opportunity to lash out at McClellan through his loyal friend, Porter, Porter was found guilty and discharged from the army on January 21, 1863. Lincoln's presidential secretary, John Hay, summed up the chain of events, saying that Porter was "the most magnificent soldier in the Army of the Potomac, ruined by his devotion to McClellan." Sixteen years later, a review board would exonerate Porter on all charges (Fleek 55; Warner 378).

WILLIAM FARRAH SMITH (1824–1903)

Smith, an 1845 West Point graduate, served in the topographical engineers and taught mathematics and surveying at West Point. He was nicknamed "Baldy" and was one of McClellan's closest friends (Catton, *Army of the Potomac* 155). On July 16, 1861, Smith was appointed colonel of the Third Vermont Infantry. He soon received command of the Vermont Brigade and promotion to brigadier general.

During the Peninsula Campaign, Smith commanded a division in the Fourth Corps. His division led the advance from Newport News Point to the Warwick-Yorktown line and met heavy resistance at the battle of Lee's Mill on April 5, 1862. Smith subsequently probed the Confederate lines and directed the attack against Dam no. 1 on April 16, 1862 (Faust 699; Boatner 775–76). Smith's artillery silenced two of the three Confederate guns covering the dam, and the Federals crossed the ford and broke the Confederate line. At this point, McClellan could have sent the bulk of his army on through the gap toward Richmond, leaving Yorktown isolated in the rear to be taken at leisure. Instead, McClellan contented himself with having captured good riverside positions for his artillery and let the opportunity pass. For his part, Smith reported, "The moment I found resistance serious and the numbers opposed great, I acted in obedience to the warning instructions of the general-in-chief, and withdrew the small number of troops exposed" (Bailey 102).

EDWIN McMASTERS STANTON (1814–1869)

Stanton studied law at Kenyon College in Ohio and later served as counsel for Pennsylvania. He gained a national reputation in defending future Federal general Daniel Sickles with the plea of temporary insanity for the shooting death of his wife's lover. In addition, Stanton served as attorney general in the last months of James Buchanan's administration. On January 13, 1862, President Lincoln named Stanton to replace the inefficient and corrupt Simon Cameron as secretary of war. The appointment would have far-reaching effects for McClellan, who would identify Stanton's arrival as the beginning of the general's difficulties with the administration (McClellan 163).

Stanton was financially honest, extremely energetic, and industrious, but he was also arbitrary, excitable, and chronically dishonest in his dealings with people. He was completely unschooled in military science, and he responded by hating and distrusting professional military officers. He quickly associated himself with the rabid Radicals of the Joint Committee on the Conduct of the War, and although he was supposed to be a friend of McClellan's, Stanton would become the general's most relentless enemy, even trying to remove him from command of the Army of the Potomac in early March (Faust 712–13; Boatner 792; Williams *Lincoln* 57; Hassler, *Commanders* 35–40).

EDWIN VOSE SUMNER (1797–1863)

Sumner held the distinction of being the oldest corps commander in either army. In 1819, Sumner was commissioned a second lieutenant in the Second U.S. Infantry and transferred to the First U.S. Dragoons four years later. Sumner fought Indians on the frontier and advanced to the rank of major. He received two brevets for gallant service during the Mexican War. Colonel Sumner served as governor of the New Mexico Territory and later commanded Fort Leavenworth. On May 16, 1861, he was appointed a brigadier general. During his forty-two years of service, he earned the nickname "Bull" for his booming voice, which could be heard over the noise of battle (Faust 842; Boatner 818).

Sumner had voted against McClellan's Urbanna Plan at the March 8 council of war, but Lincoln's reorganization of the Army of the Potomac resulted in Sumner's leading the Second Corps during the Peninsula Campaign. He proved to be a stubborn commander who was unable to coordinate his troops in battle. Indeed, Stephen Sears writes that no one in the Federal army "had risen in rank further beyond his capacity." McClellan felt that "unfortunately nature had limited [Sumner's] capacity to a very narrow extent." The Comte de Paris, a French nobleman serving with McClellan's headquarters, wrote that Sumner "has an air of stupidity that perfectly expresses his mental state" (Sears, *Gates* 71). At Williamsburg, Sumner found himself in a confused situation that these limited capabilities could not handle. He would fail miserably.

JOHN ELLIS WOOL (1783–1869)

Wool fought in the War of 1812 and rose to the rank of lieutenant colonel. He served as the inspector general of the U.S. Army until his promotion to brigadier general in 1841. Wool served in the Mexican War and received a brevet promotion to major general. In August 1861, he received command of the Department of Virginia at Fort Monroe. As the senior brigadier general, Wool resented McClellan's promotion to major general and appointment as general in chief. Lincoln's reorganization kept Wool's ten thousand troops separate from the Army of the Potomac after McClellan arrived on the Peninsula. McClellan argued with the Lincoln administration for control of Wool's men, but Lincoln and Secretary of War Stanton maintained the separation until June 6, 1862, when Wool was replaced by Major General John Dix and Dix's men were placed under McClellan's command. Wool had been promoted to major general on May 16, and he moved on to command the Middle Department (Faust 842; Boatner 948).

THE KEY CONFEDERATES
JEFFERSON FINIS DAVIS (1808–1889)

Davis, the future Confederate president, ironically had a distinguished career in the service of the United States. He graduated from West Point in 1828 and served in the dragoons for seven years. After leaving the army, Davis purchased a plantation in Mississippi and entered national politics. Elected to the House of Representatives in 1845, he resigned one year later to fight in the Mexican War. During the battle of Buena Vista, Colonel Davis's First Mississippi Rifles halted a Mexican cavalry charge and prevented an American defeat. He later served in the U.S. Senate and as secretary of war for President Franklin Pierce.

In February 1861, Davis was elected president of the Confederate States of America. Despite his political and military experience, Davis proved to be a much less effective commander in chief than did his seemingly less qualified Federal counterpart, Abraham Lincoln. Perhaps the problem was that Davis was too comfortable with his qualifications. Clifford Dowdey writes that "Davis was his own Secretary of War as well as Chief of Staff and Commander in Chief. Indeed, as everything about the military fascinated him, the President performed tasks that belonged properly to clerks in the War Office, and even in the Adjutant General's office. Conversely, as he squandered his time and energies in the field of his interests, Davis neglected affairs which properly belonged in the President's office" (*Land* 128).

Davis frequently argued with his military commanders and went through six secretaries of war in four years. He had a volatile relationship with Joseph Johnston, and the two disagreed over how best to respond to McClellan's attack on the Peninsula. Davis presumably was not disappointed when he had a chance to relieve Johnston after he was wounded at Seven Pines. With Robert E. Lee, Davis developed a much more open and collaborative relationship (Faust 208; Boatner 225–26). Dowdey, however, credits Lee with this achievement. Citing Lee's "gentle self-effacement . . . amiable disposition . . . awesome self-discipline [and] true Christian humility," Dowdey believes that "it is probable that these qualities caused [Lee] to suffer Davis' unconscious rudeness for the sake of their country" (*Land* 127).

JUBAL ANDERSON EARLY (1816–1894)

Early graduated eighteenth in West Point's 1837 class and received a commission in the Third U.S. Artillery. Second Lieutenant Early fought in the Second Seminole War but resigned his commission a year later to pursue a law career. Despite voting against secession, Early was eager to defend his state and was commissioned colonel of the

Twenty-fourth Virginia Infantry. He served as a brigade commander at First Manassas and was subsequently promoted to brigadier general. During the Peninsula Campaign, Early commanded a brigade in Daniel Harvey Hill's division. His brigade unsuccessfully charged Winfield Scott Hancock's position at Williamsburg, and Early was wounded in the shoulder. This action also initiated a lifelong feud between Early and Longstreet. Indeed, at Williamsburg, Longstreet told Hill that Early's brigade "is not in safe hands" (Eckenrode and Conrad 37). Freeman's assessment of Early after Williamsburg is that Early was "brave, ambitious to win renown, impetuous and possibly restless" (*Lee's Lieutenants* 1:192). Early would later command a brigade under Stonewall Jackson at Malvern Hill (Faust 233; Boatner 254–55).

AMBROSE POWELL HILL (1825–1865)

Hill graduated fifteenth in the West Point class of 1847. Prior to the Civil War, he served in the Mexican War, the Seminole Wars, and on the frontier. In 1856, he proposed to Ellen Marcy, who accepted; however, her father thought she could do better than Hill and forbade the marriage. Marcy ended up marrying George McClellan.

Hill was appointed brigadier general in the Confederate States Army on February 26, 1862, and commanded a brigade at Williamsburg. Douglas Southall Freeman assesses Hill's performance there as "capable, hard-hitting and skillful in control of his men" (*Lee's Lieutenants* 1:192). He was promoted to major general on May 26. Leading his division at Mechanicsville, Hill grew restless waiting for Stonewall Jackson and conducted a series of unsupported frontal attacks against the formidable Federal position. At Gaines' Mill, his five hour slugfest, which Clifford Dowdey calls "one of the longest, hardest, most unsung actions of the war" (*Seven Days* 221), paved the way for John Hood's ultimately successful assault (Dowdey, *Seven Days* 243). Freeman notes that at Glendale, Hill again "had borne the brunt of the fighting" (*Lee's Lieutenants* 1:590). Indeed, Hill emerged from the Seven Days with a reputation as a combative, if not reckless, fighter. When the *Richmond Examiner* gave Hill much of the credit for the Seven Days that James Longstreet thought he had earned, a quarrel developed between the two men. Longstreet placed Hill under arrest, and there was talk of a duel.

Robert E. Lee tactfully defused the situation by transferring Hill to Jackson's command (Freeman, *Lee's Lieutenants* 1:664–68).

DANIEL HARVEY HILL (1821–1889)

Hill graduated twenty-eighth in the West Point class of 1842 and fought in the Mexican War. Hill subsequently resigned his commission and taught at several colleges before receiving an appointment to the North Carolina Military Institute. In 1861, he was commissioned colonel of the First North Carolina Infantry and fought in the battle of Big Bethel. Hill advanced to brigadier general on June 10, 1861, and to major general on March 26, 1862 (Faust 361–62; Boatner 401).

Hill commanded a division on the Peninsula. He was extremely critical of his fellow generals—"ceaselessly critical" according to Freeman (*Lee's Lieutenants* 1:630). Robert Toombs, Theophilus Holmes, Stonewall Jackson, William Whiting, and even Robert E. Lee felt the sting of Hill's pen. In Freeman's assessment, "More often than not, Hill was sound in his adverse military judgment, though apt to disregard practical difficulties; but his insistence on pointing out the errors of others, at the same time that he dwelt on the accomplishments of his own men, irritated some of his comrades" (*Lee's Lieutenants* 1:631).

THEOPHILUS HOLMES (1804–1880)

Holmes graduated forty-fourth among the forty-six members of the West Point class of 1829 and served in the Seminole and Mexican Wars. He resigned from the army on April 22, 1861, and was subsequently appointed brigadier general in the Confederate States Army on June 5 by his West Point classmate, Jefferson Davis (Boatner 406). At the outset of the Peninsula Campaign, Holmes was the commander of the Department of North Carolina. With Lee's reconcentration of forces, Holmes brought a division to the Peninsula. Clifford Dowdey is unsparing in his contempt for Holmes, calling him an "incompetent legacy from the Old Army" and "one of the Old Army relics whom time had unfitted for war" (*Seven Days* 125, 296). Aged beyond his fifty-seven years and quite deaf, Holmes was almost comical in his ineptitude. At Malvern Hill his troops were subjected to a murderous fire from the Federal gunboats with explosions that D. H. Hill likened to "that of a small volcano." As his men scattered like quail in the bombardment, Holmes emerged from a small hut he had occupied and, cupping his hand behind his right ear, innocently remarked, "I thought I heard firing" (Wheeler 331).

JOHN BELL HOOD (1831–1879)

Hood graduated forty-fourth out of the fifty-two men in the West Point class of 1853. He was wounded in Indian fighting before resigning from the army in April 1861. He was commissioned as a lieutenant in the Confederate States Army and initially commanded Magruder's cavalry at Yorktown. On March 6, 1862, Hood was appointed brigadier general and given command of the Texas Brigade. The highlight of Hood's service during the Peninsula Campaign was at Gaines' Mill, where he launched a desperate final charge that broke the Federal line. Clifford Dowdey notes that "from that day on [Hood's Texans became] Lee's favorite shock troops" (*Seven Days* 243).

BENJAMIN HUGER (1805–1877)

Huger graduated eighth in the West Point class of 1825 and received a commission in the artillery. He served in the topographical engineers for three years before transferring to the ordnance department. During his career, Huger commanded the armory at Fort Monroe and served on the ordnance board. In the Mexican War, Huger served as Winfield Scott's chief of ordnance and received a brevet promotion to colonel. Thereafter, Huger commanded various arsenals until his resignation in 1861.

Entering the Confederate army, Huger was commissioned a brigadier general on June 17, 1861, and was promoted to major general on October 7, 1861. He held the important command of the Department of Norfolk, which contained the Gosport Navy Yard and the CSS *Virginia*.

After Johnston pulled back toward Richmond, Huger evacuated his troops, and the Federal army occupied Norfolk on May 10, 1862. The next day, the crew scuttled the *Virginia,* which opened the James River to the Federal fleet.

Huger later commanded a division at Seven Pines and in the Seven Days battles (Faust 374; Boatner 416). His painfully slow and cautious pursuit at Glendale contributed to the Federals' escape, and he is second only to Magruder as the most criticized Confederate general on the Peninsula (Freeman, *Lee's Lieutenants* 1:611). Stephen Sears describes Huger as "at heart a headquarters bureaucrat, devoid of imagination and enterprise and even energy" (*Gates* 284). It became obvious that Huger had been promoted beyond his capabilities as a field commander, and he was eventually reassigned as an ordnance inspector.

THOMAS JONATHAN JACKSON (1824–1863)

Jackson graduated seventeenth in the West Point class of 1846. He was commissioned in the artillery and served with many of his classmates in the Mexican War. Assigned to Captain John Magruder's battery, Lieutenant Jackson fought with distinction at Chapultepec and was brevetted to major. He resigned his commission in 1851 and joined the faculty at the Virginia Military Institute. Professor Jackson taught natural philosophy (similar to today's physics) and artillery. He commanded the VMI Cadet Corps at John Brown's hanging on December 2, 1859.

Jackson received a colonel's commission on April 21, 1861, and took command at Harper's Ferry. He was soon promoted to brigadier general and served with Joseph Johnston in the Shenandoah Valley and First Manassas. During the battle, Jackson's brigade held the line at the Henry House Hill and received the sobriquet "Stonewall" from Brigadier General Barnard Bee. Jackson advanced to major general on October 7, 1861, and received command of the Shenandoah Valley district (Faust 391; Boatner 432–33).

As it turned out, Jackson's greatest contribution to the Peninsula Campaign occurred not on the Peninsula but from this position in the Valley. By marching his "foot cavalry" some 170 miles in two weeks, Jackson was able to route a total of 12,500 Federals and occupy the attention of some 60,000 more (Hattaway and Jones 176). After Kernstown, a battle that amounted to a tactical defeat for Jackson, Lincoln felt so threatened that on April 3 he ordered McDowell's 30,000-man corps to remain in the vicinity of Washington rather than joining McClellan as planned. Bevin Alexander writes that Jackson "had forced a major change in Northern strategy. Seldom in the history of warfare has so small a military force achieved so enormous a strategic gain" (47). Then, by a skillful and deceptive march from Conrad's Store to Staunton, Jackson prevented the Federal armies from combining and froze McDowell in place at Fredericksburg. Jackson's biographer, James I. Robertson, notes that "what Jackson had done in the valley of Virginia changed the whole face of the Civil War in the state" (446).

Under these new circumstances, Lee planned a bold turning movement. He would fix McClellan in front while Jackson would "sweep down north of the Chickahominy, cut up McClellan's communications and rear." Jackson would further take advantage of the Confederacy's interior lines and move by rail (Hattaway and Jones 192–93).

However, the price of Jackson's excellent performance in the Valley would ultimately be paid on the Peninsula. The demanding Valley Campaign left Jackson exhausted, and he would be uncharacteristically slow and unimaginative in effecting Lee's planned concentration of forces. The Peninsula would prove that even the Mighty Stonewall "was a human being with human limitations" (J. Robertson 505).

JOSEPH EGGLESTON JOHNSTON (1807–1891)

Johnston graduated thirteenth in the West Point class of 1829. He earned a reputation for reckless bravery while fighting in the Black Hawk War, the Second Seminole War, and the Mexican War. Indeed, Captain Johnston was wounded five times and earned three brevets while in Mexico. In the 1850s, Johnston held various assignments including the chief of topographical engineers in Texas and lieutenant colonel of the First U.S. Cavalry. By the time of the secession crisis, Johnston had attained the staff rank of brigadier general and had resigned as the quartermaster general of the U.S. Army.

On May 14, 1861, Johnston was appointed a brigadier general in the Confederate States Army. He later commanded at First Manassas and received the rank of full general (Faust 400–401; Boatner 441). However, the letter President Davis sent to the Senate requesting confirmation of the nominations listed Johnston fourth, following Samuel Cooper, Albert Sidney Johnston, and Robert E. Lee. This order infuriated Joe Johnston, and from that day on he had a quarrelsome and difficult relationship with Davis (Freeman, *Lee's Lieutenants* 1:113). As Douglas Southall Freeman describes it, "Johnston doubted Davis' confidence in him; Davis doubted whether Johnston was of a temperament to justify him in trusting to the General without reservation the entire conduct of a campaign on which the fate of the Confederacy manifestly depended" (*Lee's Lieutenants* 1:203).

This friction between Johnston and Davis was apparent in their conflicting strategic views of how best to respond to McClellan's attack. The defensive-minded Johnston was dissatisfied with Magruder's Yorktown-Warwick line and advocated concentrating all available troops and fighting the Federals in front of Richmond. Davis, concurring with the advice of Robert E. Lee, ordered Johnston to attempt to hold the lower Peninsula and Norfolk, fighting the battle as far forward of Richmond as possible (Thomas, *Lee* 220–21).

Johnston and Davis never established effective communications in spite of Lee's efforts to act as a facilitator. Emory Thomas concludes that with Johnston in command, "Davis never seemed to know anything consequential until after it had happened" (*Lee* 224). When Johnston was wounded at Seven Pines, Davis appointed Lee to replace him. Johnston recognized "The shot that struck me down is the very best that has been fired in the Southern cause yet" (Sears, *Gates* 154).

ROBERT EDWARD LEE (1807–1870)

Lee graduated second in the West Point class of 1829, without a demerit, and served in the Corps of Engineers. He was promoted to captain in 1838 and participated in several civil and military engineering projects before serving on Winfield Scott's staff in the Mexican War. Captain Lee earned three brevets for gallantry and the admiration of General Scott. Before the Civil War, Lee served as the superintendent of West Point and the lieutenant colonel of the Second U.S. Cavalry. In addition, Brevet Colonel Lee quelled John Brown's Raid. After Virginia seceded from the Union, Lee resigned his commission and took command of the Virginia land and naval forces until they were incorporated into the Confederate States military. He was subsequently appointed a brigadier general in the Confederate army and attained the rank of full general on August 7, 1861. After an unsuccessful campaign in western Virginia and supervising coastal defenses in Georgia and the Carolinas, General Lee was named the presidential military adviser to Jefferson Davis (Faust 429–31; Boatner 476–77).

In this capacity, Lee attempted to defuse the volatile relationship between Johnston and Davis and had some limited success in improving communication between the two (Thomas, *Lee* 224). Also in this capacity as military adviser, Lee affected the reconcentration of Confederate troops that would counter McClellan's initially overwhelming numerical superiority on the Peninsula and began intimating to Stonewall Jackson the role he could play in the Valley to relieve Federal pressure on the Peninsula.

Lee replaced Johnston as field commander on May 31, 1862, after Johnston was wounded at Seven Pines. This event changed the course of the war. In contrast to the strained relationship between Johnston and Davis, Lee and Davis enjoyed a smooth-working communication and collaboration. Even more significantly, Lee exchanged Johnston's largely defensive strategy for an audacious offensive-defensive strategy that would seize the initiative and dictate the timing and tempo of operations. In Lee's mind, "It is only by concentration of our troops that we can hope to win any decisive advantage. . . . [We] must decide between the positive loss of inactivity and the risk of action" (Weigley 102).

Lee's greatest challenge on the Peninsula was in getting his brilliant but complex plans to be properly executed. Failed by individual commanders and with little help from his staff, Lee struggled to execute turning movements that would allow him to achieve local numerical superiority. In spite of these difficulties, what turned the tide on the Virginia Peninsula during the late spring and early summer of 1862 was generalship. Robert E. Lee and George McClellan were simply in different leagues. As Bruce Catton summarizes the campaign, "The soldier who had fought with all the odds against him had taken hair-raising risks and had won; the soldier who had had all of the advantages had refused to risk anything and lost; and now the last chance that this ruinous war could be a relatively short one was gone forever" (*Hallowed Ground* 144).

JAMES LONGSTREET (1821–1904)

Longstreet, an 1842 graduate of West Point, served in Florida and Mexico. Lieutenant Longstreet was wounded at Chapultepec and brevetted to major. He resigned his commission on June 1, 1861, and was appointed a brigadier general in the Confederate States Army sixteen days later. A veteran of First Manassas, Longstreet was promoted to major general on October 7, 1861, and commanded a division under General Joseph E. Johnston. During the May 5, 1862, battle of Williamsburg, Longstreet fought an excellent rearguard action that allowed the Confederate artillery and supply train to escape toward Richmond.

Highly skilled in the conduct of the defense, Longstreet was not as effective in offensive situations. At Seven Pines, he took the wrong road, arrived late on the battlefield, and was excessively passive in command (Faust 444–45; Boatner 490–91). He would, however, show improvement during the Seven Days. His biographers, H. J. Eckenrode and Bryan Conrad, summarize that Longstreet's "offensive tactics at Seven Pines had been wretched. At Gaines' Mill he had handled his men to far better effect, and at Glendale he had conducted a severe offensive action with considerable success. But it still remained true of him that he was essentially a defensive fighter, not offensive, and he was to continue thus to the end of the war" (86).

JOHN BANKHEAD MAGRUDER (1807–1871)

Magruder graduated from West Point with the "somewhat undistinguished" class of 1830 (Freeman, *Lee's Lieutenants* 1:14–15) and was commissioned in the artillery. His idiosyncrasies and intense personality make him one of the most interesting characters on the Peninsula.

In the Mexican War, Magruder was a captain commanding an independent battery that would follow the flow of the battle. Magruder had a reputation as a fighter, and Lieutenant Thomas Jackson worked hard to join his command. Jackson told a friend, "I wanted to see active service. I wished to be near the enemy and in the fight, and when I heard John Magruder had got his battery, I bent all my energies to be with him, for I knew if there was any fighting to be done, Magruder would be on hand" (J. Robertson 61).

Magruder initially did not seem to return such a high opinion of Jackson. Major General John Tidball wrote in his memoirs that "Prince John Magruder, as he was called because of his affected elegance, was in reality only a prince of humbugs. No greater difference could possibly exist between men than between Magruder and [Jackson]. The former was a dissolute, reckless braggart, whose chief aim in life was display and deceit. He despised Jackson because first, although the latter was from Virginia, he was not an F.F.V. [First Family of Virginia] and did not, like himself, pretend to be one and put on airs accordingly; and secondly, [Magruder] saw nothing in [Jackson] to admire except what he considered only 'stupid bravery'" (J. Robertson 60–61). However, after Jackson's sterling performance at Contreras, Magruder's opinion seemed to change, considering Jackson's behavior "conspicuous throughout the whole day" and adding that "I cannot too highly commend him to [Major General Daniel Twiggs's] favourable consideration" (J. Robertson 64).

Then, at Chapultepec, Magruder and Jackson worked side by side, manhandling guns under heavy fire in one of the day's most decisive small unit actions (J. Robertson 67). Brigadier General Gideon Pillow stated that "Captain Magruder's field battery, one section of which was served with great gallantry by himself, and the other by his brave Lieut. Jackson, in the face of a galling fire from the enemy's position, did invaluable service preparatory to the general assault" (J. Robertson 69). This was Magruder at his best—hands-on, full of intensity and bravery, and in charge of an operation the size of which he could personally manage. Magruder was brevetted to lieutenant colonel for his conduct in Mexico.

Magruder and Jackson would again find their destinies linked during the Seven Days. On June 29, Lee intended for Jackson to cross the Chickahominy, join Magruder, and vigorously pursue the retreating Federals. Unfortunately, the two could not replicate their Mexican War success. Jackson would be uncharacteristically slow, and Magruder "seemed to fall to pieces step by step" (J. Robertson 489). Historians continue to debate the reasons for Jackson's shortcomings during the Seven Days. Magruder's performance is simpler to understand. He had been endowed with responsibilities beyond his level of ability.

Magruder had been promoted to brigadier general on June 17, 1861, and reached the rank of major general on October 7, 1861, primarily on the strength of his June 10 victory at Big Bethel, northwest of Newport News. Major General Benjamin Butler had attacked one of Magruder's outposts, and in the resulting confusion, the Federals withdrew back to Fort Monroe. Only seventy-six Federal and eight Confederate casualties occurred in what amounted to a twenty-minute encounter. Douglas Southall Freeman observes that "three years later, such a clash would have been accounted a skirmish and, if reported at all, would have been the subject of a two-line dispatch. It was different on June 10, 1861. . . . Forthwith [Magruder] took his place among the foremost of Southern celebrities, a hero second only to Beauregard in the esteem of the Confederacy" (*Lee's Lieutenants* 1:18–19).

Big Bethel was indeed a small action. Not more than three hundred of the fourteen hundred Confederates had been engaged simultaneously. Now Magruder found himself commanding the Army of the Peninsula and constructing successive defensive lines to prepare against a massive Federal attack. Such a dispersed responsibility would be a challenge for someone who, as Freeman notes, had "an excited, overzealous desire to do all his work in person" (*Lee's Lieutenants* 1:xxxiv).

Magruder's artillery training would come in handy in locating the batteries that would cover the Federal avenues of approach, but laying out the defensive lines would be a greater challenge. By his own admission, Magruder was not much of an engineer (Foote 400). Nonetheless, on the Peninsula, the terrain would favor the defense.

Magruder might have lacked engineering skills and might have had more responsibilities than he could handle, but he did use the talents he had. Among these was experience as an amateur actor and a tremendous flair for the dramatic. Outnumbered on the Peninsula, he would use his theatrical skill to orchestrate brilliant deceptions. Shelby Foote observes that "If Magruder was high-strung and overimaginative by ordinary standards, it presently developed that these qualities, so doubtful in a military leader, could be positive advantages in an extraordinary situation, such as the one that involved him now" (399). Magruder would perform disappointingly during the Seven Days, but his actions earlier in the campaign had done much to make the Seven Days possible.

GABRIEL JAMES RAINS (1803–1881)

Rains graduated thirteenth in the West Point class of 1827 and fought in the Second Seminole War and the Mexican War. On July

31, 1861, Rains resigned as the lieutenant colonel of the Fifth U.S. Infantry. He was subsequently appointed a brigadier general and commanded a brigade during the Peninsula Campaign. Dowdey describes Rains as "an old-time army man unsuited for combat" (*Seven Days* 96), and D. H. Hill criticized Rains for not delivering a second flank attack at Seven Pines (Freeman, *Lee's Lieutenants* 1:268). Freeman identifies the problem, saying, "Rains was at heart a scientist, and was more interested in explosives than in field command" (*Lee's Lieutenants* 1:268).

Indeed, when the Confederates evacuated Yorktown, Rains pioneered the use of land mines (subterranean torpedoes) by burying artillery shells along the roads and beach. Both Federal and Confederate commanders criticized this "barbaric" method of warfare, and on June 18, Rains was assigned to the river defenses, where the use of his torpedoes was "clearly admissible" (Freeman, *Lee's Lieutenants* 1:269). Along with ironclads and balloons, Rains's use of mines underscores the technological advances employed during the Peninsula Campaign (Faust 610–11; Boatner 676; Longstreet 79).

GEORGE WYTHE RANDOLPH (1818–1867)

Randolph, a grandson of Thomas Jefferson, graduated from the University of Virginia and served in the U.S. Navy. In 1850, Randolph established a law practice in Richmond and became a civic leader. Moreover, he organized a volunteer artillery company, the Richmond Howitzers, and participated in the execution of John Brown. In 1861, Randolph served as a delegate in the state secession convention and commanded the Richmond Howitzers stationed on the Peninsula. On June 10, 1861, Colonel John Magruder's infantry, supported by Randolph's artillery, repelled the Federal attack at Big Bethel.

On February 22, 1862, Randolph was promoted to brigadier general, and on March 22, 1862 he was appointed Confederate secretary of war. Magruder constantly pleaded with his former subordinate for more weapons and supplies for the Peninsula's defense. Indeed, as secretary of war, Randolph found himself much more involved with logistical concerns such as these than with operational matters.

Perhaps Randolph's most significant contribution was persuading Jefferson Davis to sign the Conscription Act, the first in American history, to prevent the dissolution of the Confederate Army after the one-year enlistments expired in April 1862 (Faust 613–14; Boatner 678–79). Beyond that, Randolph failed to work well with Davis, who was much more comfortable seeking advice from Lee. Randolph, whose stamina was already sapped by tuberculosis, was frustrated by Davis's frequent and sometimes protracted discussions of military matters, which often seemed to lack any sense of urgency toward reaching a decision (Jones 78). Randolph resigned as secretary of war on November 15, 1862, to resume a field command (Boatner 679).

GUSTAVUS WOODSON SMITH (1821–1896)

Smith graduated from West Point eighth in the class of 1842 and was commissioned in the engineers. He provided capable but limited service in the Mexican War, where he served as second in command of the engineering company to which George McClellan was assigned. Smith resigned from the army in 1854 to pursue a career in civil engineering. In this capacity, he advanced to the position of commissioner of streets in New York City.

Smith was born in Kentucky, and, like his native state, he initially chose a neutral course early in the Civil War. After five months, he declared himself for the Confederacy and reported to Richmond. He received an appointment as a major general in spite of his rather limited qualifications. No other man except Albert Sidney Johnston and Leonidas Polk had previously received so high an initial commission (Freeman, *Lee's Lieutenants* 1:162). Smith would not live up to these expectations.

Smith was personally courageous, but Stephen Sears writes that "simply the thought of the responsibilities of high command seemed to unnerve" the general (*Gates* 140). On the Peninsula, this hesitancy would work to the Confederacy's advantage. Smith was Johnston's second in command, and when Johnston was wounded, Smith succeeded him. However, when President Davis met with Smith to ascertain the situation, Smith was unimpressive, telling Davis that at the time the general "could not determine, understandingly, what was best to be done." By Smith's own admission, "Mr. Davis did not seem pleased with what I said" (Sears, *Gates* 141). On June 1, Robert E. Lee rode up to Smith's headquarters with Davis's instructions to replace Smith and assume command of the Army of Northern Virginia. Smith would be spared the responsibility with which he was so uncomfortable, and the Confederacy would be much better for it.

JAMES EWELL BROWN "JEB" STUART (1833–1864)

Stuart graduated thirteenth in the 1850 class at West Point and received a commission in the cavalry. Lieutenant Stuart was stationed in Texas and Kansas before serving as a volunteer aide to Colonel Robert E. Lee during John Brown's Raid. On May 10, 1861, Stuart resigned his captain's commission and two weeks later received a commission as colonel of Confederate cavalry.

He organized the First Virginia Cavalry and fought at First Manassas. Promoted to brigadier general on September 24, 1861, Stuart commanded a cavalry brigade during the defense of the Warwick-Yorktown line. His cavalry covered the retreat to Williamsburg on May 4, 1862, and held back the Federal cavalry and infantry until Confederate infantry reoccupied the Williamsburg line. From June 12 to June 15, Stuart rode entirely around McClellan's army, informing Lee that the Federal right flank was vulnerable. It was a role for which the adventurous Stuart was well suited (Faust 727–28; Boatner 812–13; Thomas, *Bold Dragoon* 5–7).

In such actions, Stuart's cavalry provided the Confederates with an excellent reconnaissance capability and gave Lee a marked intelligence advantage over McClellan. Joe Johnston assessed, "I know of no one more competent than [Stuart] to estimate the occurrences before him at their true value" (Sears, *Gates* 167).

JOINT OPERATIONS AT THE OUTSET OF THE CIVIL WAR

The personalities of these and other Federal and Confederate leaders would be critical in the Peninsula Campaign because of the absence of joint doctrine. *Doctrine* represents "the fundamental principles by which the military forces or elements thereof guide their actions in support of national objectives. It is authoritative but requires judgment in application" (FM 101-5-1 1-55). *Joint* means involving two or more of the military departments, such as the army and the navy. Joint doctrine was conspicuously absent at the time of the Civil War, so commanders would be forced to work out their issues largely on the basis of their individual personalities.

From the very beginning of the Civil War, Federal strategy recognized that success would come from the work of both the army and the navy. General in Chief Winfield Scott proposed an "Anaconda Plan" that involved two parts: a tight blockade of the Confederacy's ports and then a move of an army of sixty thousand men, accompanied by gunboats, down the Mississippi River to seize and hold it from Cairo, Illinois, to the Gulf of Mexico.

The Anaconda Plan, however, was more a diplomatic policy than a plan for strategic military action. Moreover, this policy would take years to make its effects felt. President Lincoln could not adopt such a time-intensive and diplomatic-focused course (Williams, *Lincoln* 16–18). Nonetheless, the Anaconda Plan did represent a certain level of joint strategic thought.

One key reality that hindered more detailed joint strategic thinking was the absence of anything to parallel today's Joint Chiefs of Staff. The army and navy developed boards to help cope with the demands of large-scale war, coordinate the various service bureaus, and provide the president with professional military advice on operational questions (Hattaway and Jones 102), and these boards made some forays into joint issues. The War Board was headed by Major General Ethan Allen Hitchcock, and its members were all from the army (Hattaway and Jones 121). Nonetheless, the board discussed several seemingly joint issues, including the threat posed by the Confederate ironclad *Virginia,* sea-coast defense, ironclad vessels, and amphibious expeditions (Hattaway and Jones 124).

However, the limitations of officers of purely army backgrounds trying to wrestle with such joint topics were often apparent. When Secretary of War Edwin Stanton asked the board how to destroy the *Virginia,* board member Joseph Totten replied, "We are not sailors, Mr. Secretary." Likewise, Montgomery Meigs concluded, "this task must be left to the Navy" (Hattaway and Jones 133).

Captain Samuel DuPont directed the Navy Board (also called the Blockade Board, the Strategy Board, and the Committee on Conference)

(Reed 8), which included an army representative, Major General John Barnard of the Corps of Engineers (Hattaway and Jones 135). Indeed, the army engineers provided the navy with valuable topographical information that assisted with the navy's blockade strategy (Niven 357). While these boards represented significant developments in the efficiency of the respective services, they were not intended to be, nor did they become, anything like a joint staff. In fact, General in Chief Scott did not even have a counterpart commander of the navy with whom to coordinate (Reed xiii). This situation was exacerbated by the fact that Secretary of the Navy Gideon Welles and Secretary of War Stanton did not get along. Moreover, Welles continually complained that the army did not keep him informed of its plans (Niven 480–81).

In addition, no national-level joint staff existed to develop joint doctrine or advise the president or joint task forces at the operational level to plan and command joint operations. Indeed, T. Harry Williams notes that "the inadequacies of the command system, which was not equipped in 1862 to manage a complicated operation calling for common action by two branches of the service," was one obstacle to the Peninsula Campaign's achieving true jointness (*Lincoln* 76). Thus, the planning and execution of joint operations depended completely on ad hoc actions by the responsible commanders. As Scott Stuckey observes, "Neither command arrangements nor doctrine for joint operations existed at the time [of the Civil War]. Successful joint operations, like much else, would have to be improvised by those on the scene" (98–99).

Key to the difference between success and failure was the commanders' ability to achieve unity of effort. Unity of effort focuses on cooperation rather than command. Accordingly, it is distinct from the traditional principle of war of unity of *command,* which requires "that all forces operate under a single commander with the requisite authority to direct all forces employed in pursuit of a common purpose." Unity of *effort,* conversely, "requires coordination and cooperation among all forces toward a commonly recognized objective, although they are not necessarily part of the same command structure" (Joint Pub 3-0, A-2). Joint operations during the Civil War required unity of effort. To get it, effective communication, personal relationship skills, consensus building, and shared purpose would all be required. On the Peninsula, McClellan and Goldsborough would not achieve this.

CHAPTER 2

OPENING MOVES

Shortly after replacing Winfield Scott as general in chief of the Federal army on October 31, 1861, McClellan met Colonel Rush Hawkins of the Ninth New York, who was making a report to the cabinet. Hawkins had just returned from a successful amphibious operation against Hatteras Inlet, North Carolina, in which his command had cooperated with naval forces under Admiral Silas Stringham. Employing a landing party of 319 men, the Federals had suffered just one casualty while capturing 670 prisoners and thirty-five cannon. The contribution of the navy had been significant, so much in fact that one historian concludes that the Confederate surrender "resulted almost entirely from the effectiveness of Stringham's bombardment" (Boatner 385). Indeed, Rowena Reed concludes that the attack demonstrated to the navy that shore batteries could be reduced by naval bombardment alone and that, in this case, the army troops "had proved more of a handicap than an asset" (14).

When the meeting ended, McClellan called Hawkins aside and began to ask him some questions. However, Hawkins soon learned that McClellan's interest was not in Hatteras Inlet at all but in the area around Norfolk and Hampton Roads, Virginia. Hawkins was eager to respond. As a matter of fact, he had already advised Major General John Wool, commander of the Federal garrison at Fort Monroe, that the army ought to conduct an amphibious landing at the tip of the Virginia Peninsula and move toward Richmond from the east. Hawkins drew McClellan a rough sketch of the terrain that indicated the road networks and showed how gunboats could be used to provide both transportation and flank protection for an invading army using the York and James Rivers.

McClellan listened enthusiastically. Hawkins's suggestion coincided with McClellan's desire to avoid a frontal assault against General Joe

Johnston's entrenched Confederates around Manassas and Centreville. McClellan pocketed Hawkins's map and began to develop the idea (Catton, *Army of the Potomac* 87).

McCLELLAN'S PLAN

McClellan was a secretive man who was immensely distrustful of politicians. Yet he knew that an amphibious operation of the scale he envisioned would require a tremendous financial outlay. Thus, in early November 1861, he approached Secretary of the Treasury Salmon Chase with the rudiments of the plan. McClellan asked Chase if he could financially support the movement, to which Chase replied, "I could get along under existing arrangements until about the middle of February." This suited McClellan perfectly, since he anticipated the entire operation being completed by February 1, 1862 (Catton, *Terrible Swift Sword* 129).

Aside from money, an additional factor that likely contributed to McClellan's decision to contact Chase was that at this time, the U.S. Coast Survey was a bureau of the Treasury Department. Coast Survey Superintendent Alexander Bache had been active in planning the Federal naval blockade of Confederate ports and was a member of the Navy Board (Reed 7; Hattaway and Jones 135). Perhaps even more importantly, Chase was McClellan's "most vocal supporter in the Cabinet" (Sears, *McClellan* 131). Nonetheless, McClellan's early decision to coordinate with Chase rather than to take President Lincoln into his full confidence would prove to be a fatal error.

McClellan's proposal would evolve into what became known as the Urbanna Plan, which was designed to be a bold, swift movement by water from Annapolis, Maryland, through the Chesapeake Bay to the mouth of the Rappahannock River. The landing site would be the small hamlet of Urbanna, which lay about sixty road miles northeast of Richmond. From there, McClellan planned "to gain Richmond by a rapid march, [but failing that] to cross the James and attack the city in the rear, with the James as a line of supply" (McClellan 163).

Such a plan had a number of attractive qualities. Foremost in McClellan's mind was the fact that the plan avoided a frontal assault against the entrenched Confederates, instead flanking the defense and forcing Johnston to evacuate his Manassas-Centreville stronghold and rush south to defend his capital. In the process, McClellan's marching distance to Richmond would be reduced by more than fifty miles. With luck, he might even reach Richmond before Johnston could. McClellan figured that he would need a force of one hundred thousand men, transported to Urbanna in two lifts of fifty thousand each, to succeed.

The roots of McClellan's plan can be traced to his Mexican War experience fifteen years earlier and General Winfield Scott's amphibious turning movement to Vera Cruz. Like Scott, McClellan enjoyed naval superiority over his foe (Hattaway and Jones 87). The fact that the Federal navy had made significant contributions to the recent victories at Hatteras Inlet, as Hawkins had described, as well as at Port Royal Sound, South Carolina, and Ship Island, Mississippi, made the plan even timelier.

McClellan's next confidant was Brigadier General John Barnard, the army engineer and the member of the Navy Board charged with establishing liaison between the army and the Navy Board (Hattaway and Jones 135). In a discussion in early December, Barnard pointed out that an amphibious move of this scale would be necessarily time-consuming. Johnston could very well take advantage of this fact to move to and capture Washington while McClellan was still at sea. Barnard felt that a force of at least a hundred thousand men would have to be left around Washington to safeguard the city. Additionally, Barnard felt that a waterborne campaign against Richmond would require the capture of Norfolk as a first step. The navy too favored an operation against Norfolk as a means of thwarting the Confederacy's efforts to develop the ironclad *Virginia*. If McClellan was going to move via Urbanna, Barnard felt that the operation against Norfolk should be initiated at once (Catton, *Terrible Swift Sword* 129–30).

In reality however, nothing was being initiated at once. It was now December, and McClellan planned to have the whole affair wrapped up by February 1, but, aside from whatever McClellan was mulling over in his mind, no real preparations had been made. President Lincoln, the man who ultimately would have to approve the plan, was entirely ignorant that such an idea even existed.

Without knowledge of McClellan's plan, Lincoln began to grow concerned over the Army of the Potomac's inactivity. Around December 1, he sent McClellan a proposal for a combined frontal and flank attack of the Confederates at Manassas, and he asked McClellan to respond with information about when such a move could be initiated and how many men it would take. McClellan gave Lincoln the information he requested but also noted that he did not approve of the plan, adding, "I have now my mind turned actively toward another plan of campaign that I do not think at all anticipated by the enemy nor by many of our own people" (Williams, *Lincoln* 52).

This mysterious hint was all that McClellan chose to reveal to his commander in chief about the Urbanna Plan. Any real action that McClellan was contemplating was further delayed, first by torrential rains and then by a three-week period in which McClellan was bedridden with typhoid fever. The situation was exacerbated by the concurrent session of the Joint Committee on the Conduct of the War, in which Radical Republicans such as Zachariah Chandler and Benjamin Wade fanned anti-McClellan sentiment (Trefousser 184–86).

Finally, Lincoln could stand no more. On January 10, he convened a White House conference with several cabinet members and two of McClellan's division commanders, Major General Irvin McDowell and Brigadier General William Franklin. Lincoln set the tone for the meeting by announcing that "If General McClellan does not want to use the Army, I would like to borrow it provided I could see how it could be made to do something" (Bailey 68).

Lincoln then turned to the generals for suggestions. McDowell argued for an overland offensive against the Confederates in northern Virginia, while Franklin proposed an amphibious venture against Richmond that approximated the Urbanna Plan. These two courses of action were debated in subsequent meetings, which McClellan eventually learned were going on behind his back. He struggled out of bed and showed up unannounced at the January 13 session.

Still McClellan remained largely silent. Chase challenged him to reveal his plans, but McClellan declined, explaining that some of those present "were incompetent to form an opinion and others incapable of keeping a secret." He did indicate that he had fixed a schedule for an advance, but he did not elaborate. Nonetheless, this seemed to placate Lincoln for the time being at least, and he adjourned the meeting. McClellan, however, had made a powerful enemy of his erstwhile benefactor, Chase, and had also irritated several other cabinet members (Bailey 68).

McClellan's respite was to be short-lived. On January 20, Edwin Stanton replaced Simon Cameron as secretary of war. Stanton was impatient and bellicose, and he and McClellan soon did not get along. According to McClellan, "Soon after Mr. Stanton became Secretary of War it became clear that, without any reason known to me, our relations had completely changed. Instead of using his new position to assist me he threw every obstacle in my way, and did all in his power to create difficulty and distrust between the President and myself" (Wheeler 65). McClellan observed that with Stanton's arrival, "The impatience of the Executive immediately became extreme, and I can attribute it only to the influence of the new Secretary, who did many things to break up the free and confidential intercourse that had heretofore existed between the President and myself" (McClellan 164). To McClellan, Stanton was "without exception the vilest man I ever knew or heard of" (Sears, *McClellan* 188).

Even allowing for McClellan's paranoia, this new influence in the cabinet undoubtedly had much to do with Lincoln's decision to issue what he called "General War Order no. 1" on January 27. The message began with instructions "that the 22nd. day of February 1862, be the day for a general movement of the Land and Naval forces of the United States against the insurgent forces." A supplementary order issued four days later specified that the Army of the Potomac would march down to cut the railroad southwest of Manassas Junction (Bailey 72).

One goal of General War Order no. 1 was undoubtedly to force McClellan to divulge his plans, and the order did just that. On February 3, McClellan sent Stanton a twenty-two-page missive outlining the merits of his Urbanna Plan and pointing out the disadvantages of Lincoln's plan to attack at Manassas.

McClellan noted that Johnston occupied a "strong central position" protected by "a strong line of defense enabling him to remain on the defensive, with a small force on one flank, while he concentrates everything on the other." An overland turning movement in the west would be difficult because of the "long line of wagon communication" needed, and a turning movement in the east would be difficult because the Confederate defenses were tied in with the natural obstacles of the Occoquan and Potomac Rivers.

McClellan felt that even if the overland attack succeeded, the Confederates merely "could fall back upon other positions and fight us again and again." While retreating, the enemy "would destroy his rail-road bridges and otherwise impede our progress" until reaching the "intrenchments of Richmond." This would disrupt the Federal supply operation and force the Army of the Potomac "to seek a shorter land route to Richmond" by the rivers southeast of the city. Thus, McClellan felt that at the conclusion of Lincoln's plan, he could very well be in no better position than he would be at the start of the Urbanna Plan. Why then the general asked, should he "spend so much more time" getting to Richmond when he could "adopt the short line at once?" (Hattaway and Jones 93).

The Urbanna Plan, conversely, "would probably cut off Magruder in the Peninsula, and enable us to occupy Richmond before it could be strongly re-enforced. Should we fail in that, we could, with the coopera-tion of the navy, cross the James and throw ourselves in the rear of Richmond, thus forcing the enemy to come out and attack us" because "his position would be untenable with us on the southern bank of the river." The threat to Richmond would force Johnston into a decisive engagement outside of his prepared positions on a battlefield selected by the Federals (Hattaway and Jones 93).

Lincoln may have already read McClellan's letter or it may have still been on Stanton's desk when on February 3 the president sent McClellan a list of five questions regarding the proposals. Lincoln was prepared to yield to McClellan if he could satisfactorily answer the following questions:

> 1st Does not your plan involve a greatly larger expenditure of time, and money, than mine?
> 2nd Wherein is a victory more certain by your plan than mine?
> 3rd Wherein is a victory more valuable by your plan than mine?
> 4th In fact, would it not be less valuable, in this, that it would break no great line of the enemy's communications, while mine would?
> 5th In the case of disaster, would not a safe retreat be more difficult by your plan than mine?

McClellan made no written reply to Lincoln's note, although it is pos-sible that he responded in person (Nevins 41–42). It is also possible that McClellan simply let his earlier letter stand as sufficient response. That let-ter was in fact the first time that McClellan had taken the administration fully into his confidence about his strategy. This novel frankness obviously was pleasing to Lincoln, and, in spite of his misgivings, he yielded to the general's wishes. However, the fact that the plan had less than Lincoln's full support is evidenced by the fact that the President waited until February 27 to issue an order to accumulate the ships necessary to make the move (Williams, *Lincoln* 65).

Key to Lincoln's hesitancy was the same concern Barnard had raised back in December—the safety of Washington. By McClellan's intelligence service's own estimate, the Confederates had a force of more than 115,500 men poised at Manassas, just thirty miles outside of the Federal capital (Nevins 42). Like Barnard, Lincoln could see the possibility of Johnston taking advantage of McClellan's absence and striking Washington. While Lincoln ostensibly approved McClellan's plan, the president did not forget his misgivings.

Lincoln's reservations manifested themselves on March 8, 1862, when he called McClellan to the White House and renewed his former objections to the Urbanna Plan. During the course of the discussion, Lincoln informed McClellan of certain rumors that were being circulated that implied that the plan "was conceived with the traitorous intent of removing its defenders from Washington, and thus giving over to the enemy the capital and the government thus left defenseless" (Foote 254). In spite of Lincoln's assurances that others, not he, thought the plan was "traitorous," McClellan was naturally offended. In response, McClellan restated all his arguments in favor of the plan and then offered to lay the entire matter before a council of war comprised of his division commanders.

Lincoln liked the idea of subjecting the plan to the scrutiny of a military council. If the collective wisdom of the generals endorsed McClellan's plan, then it must have some merit. Either way, responsibility for its success or failure would now be transferred to the military (Williams, *Lincoln* 66–67).

That same day, McClellan called his generals together and put the matter to a vote. Four of the commanders—McDowell, Samuel P. Heintzelman, Edwin V. Sumner, and Barnard—voted against it. The eight others supported it, although Erasmus D. Keyes did so only on the condition that the Potomac River batteries be first reduced. This vote of confidence was decisive, and Lincoln had to recognize the stated position of his military leaders. In a later discussion with Stanton, Lincoln rationalized,

"We can do nothing else than accept their plan and discard all others. . . . We can't reject it and adopt another without assuming all the responsibility in the case of the failure of the one we adopt" (Williams, *Lincoln* 67).

Lincoln, however, was still not ready to throw his full support behind the plan. After the council announced its decision, he issued three orders that would seriously restrict McClellan's execution of the operation.

The first of these was a reorganization of the Army of the Potomac into four corps. The question of a corps reorganization was not a new one. Lincoln, Stanton, McClellan, and the Radical leaders in Congress had discussed it for weeks. However, now what was once a technical military problem had become a political military one.

The officers of the Army of the Potomac were divided into two distinct factions. One group included several senior generals who were Republican in politics and favored wartime emancipation. Several of these officers were older than McClellan and resented his outranking them. In the other group were predominantly younger officers who were loyal to McClellan. McClellan called these men "gentlemen and Democrats" (Foote 253). He had brought them into the army and raised them to their present status.

McClellan opposed any reorganization that would strengthen the power of the Republican generals. Now, on the eve of a major campaign, Lincoln instituted a move without consulting McClellan. The twelve divisions were organized into four corps, which would be commanded by McDowell, Sumner, Heintzelman, and Keyes. At the March 8 council, McDowell, Sumner, and Heintzelman had voted against the Urbanna Plan, and Keyes had agreed to it only conditionally (Williams, *Lincoln* 68; Nevins 45–46). Men such as Franklin who had supported an amphibious plan similar to McClellan's at the January 10 meeting with Lincoln were bypassed (Foote 253). McClellan would be entering into the campaign with commanders who were not enthusiastic about his plan.

In his second order, Lincoln directed that no change could be made in the base of the Army of the Potomac without leaving in and around Washington such a force as McClellan and all the corps commanders agreed would leave the capital "entirely secure." Furthermore, no more than half the army could be moved to a new base until the enemy blockade of the lower Potomac was lifted. Finally, the move had to begin not later than March 18.

The most important of Lincoln's orders came on March 11. McClellan was relieved as general in chief and retained command only of the Army of the Potomac. Major General Henry Halleck would command all forces in the West, and Major General John Frémont would command the Mountain Department, made up of western Virginia and East Tennessee. In eastern Virginia, McClellan's would be one of four commands. Nathaniel Banks would command in the Shenandoah Valley, McDowell would command the forces between Washington and Richmond, and Wool would command the forces at Fort Monroe. No successor to McClellan was named as general in chief. Lincoln himself would take direct control of military operations (Hattaway and Jones 97).

Many observers have identified this decision as having a critically debilitating effect on the Peninsula Campaign. For example, Robert Tanner writes that with this reorganization, McClellan's "Peninsula Campaign masterplan was abandoned. It had been replaced with four separate Federal armies operating in four tight little compartments toward four different objectives. All four would be co-ordinated, if at all, by Lincoln and Stanton, men completely without military experience" (131–32).

In spite of these criticisms, this decision of overall command was fundamentally sound because it would free McClellan from distractions elsewhere and allow him to concentrate on the Urbanna operation. It is certainly not correct to trace McClellan's failure on the Peninsula to this decision. Scott Stuckey dismisses the arguments of those who see such a connection for several reasons:

First, [such an interpretation] posits that McClellan could have, with the nebulous powers of general in chief, achieved results with field armies that he was unable to do with his own when in active command. Second, the notion that McDowell's corps was essential to victory on the peninsula is nonsense. McClellan always greatly overestimated his opponents, and McDowell would not have made a difference. Third, McClellan had no authority whatsoever over naval forces. To assume that as general in chief in Washington he could have forced Army-Navy cooperation in distant theaters flies in the face of experience throughout the Civil War. Finally, this interpretation simply ignores fatal flaws in his character. An unwillingness to move quickly and fight, consistent overestimation of his opponents, secretiveness about his intentions, and contempt for his political masters in this most political of wars destroyed McClellan in the final analysis. There is absolutely no reason to think that if he had been general-in-chief and given everything he wanted in the Peninsula Campaign it would have made any difference. (94)

While the change was far from the decisive point of failure in the campaign, it nonetheless did not set well with McClellan and must be viewed as having a negative effect on his psyche. He felt that the shakeup robbed him of cooperation outside of eastern Virginia, and the fact that all principal cabinet members approved of the move had to shake his confidence (Nevins 46). To add insult to injury, McClellan learned of the change via the newspapers rather than by personal message. As historian T. Harry Williams observes, "If ever a general taking the field needed the complete trust of his superiors, McClellan did, and if ever a general lacked it, he did" (*Lincoln* 74).

These political considerations had made it a difficult, uphill battle for McClellan to get his Urbanna Plan in motion. Now, battlefield considerations were about to make the scheme obsolete before it ever really had a chance to start. The whole idea of the maneuver had been to get to Johnston's rear, force him to conduct a hasty retreat, and then make him fight to defend Richmond on terms favorable to McClellan.

Unfortunately, Johnston was not willing to cooperate with such a scheme. He was beginning to feel very vulnerable with his position at Manassas, especially since the coming warm spring weather would dry the roads and make it possible for McClellan to attack with superior numbers (Bailey 81–82). Johnston had no intention of waiting around long enough for this to happen. On March 7, he ordered all of his troops east of the Blue Ridge Mountains—some forty-two thousand effectives— to withdraw to the Rappahannock River, nearly half the distance to Richmond. Only Major General Stonewall Jackson's fifty-four hundred men would remain in the Shenandoah Valley to threaten the right flank of any Federal advance (Bailey 83).

Johnston's move completely negated the basis of the Urbanna Plan. Instead of turning the Confederates and getting between them and Richmond, McClellan now faced an enemy that had occupied the area from which he proposed to begin his operation.

By this time, however, McClellan was committed to an amphibious campaign. Back in his February 3 letter to Lincoln, the general had noted that if Urbanna did not offer a suitable landing site, he could also use Mob Jack Bay or Fort Monroe and then advance up the region between the James and York Rivers, known locally as the Peninsula. Either river could be used as a line of communication (Williams, *Lincoln* 72). Such a maneuver, however, did not offer the opportunity to cut off the Confederates as the Urbanna Plan did. Indeed, strategist B. H. Liddell Hart concludes that the plan now represented more "of a shorter direct approach to Richmond . . . than an indirect approach in the true sense" (Hart 126). Archer Jones seems to agree, concluding that "McClellan had used a turning movement to reach Richmond, but his had not involved surprise nor had it interrupted the enemy's communication" (Jones 73).

The campaign now would require a slow, toilsome march ending in a toe-to-toe fight at Richmond (Catton, *Army of the Potomac* 99). For this reason, McClellan believed the Peninsula option would result in "less celerity and brilliancy of results" than the Urbanna Plan (Nevins 47). At one point he even called Fort Monroe his "worst coming to worst" alternative to the Urbanna Plan (Sears, *McClellan* 164). Nonetheless, he greatly preferred it to the overland route. At least part of the reason must be attributed to his characteristic inflexibility and tendency to give higher priority to the plan itself than the evolving situation.

McClellan decided to put the matter before his corps commanders, and he called a council on March 13 at Fairfax Court House, Virginia. The corps commanders unanimously voted to adopt the plan as long as certain conditions could be met. The council resolved,

> That the enemy, having retreated from Manassas to Gordonsville, behind the Rappahannock and the Rapidan, it is the opinion of the Generals commanding army corps that the operations to be carried on will be best undertaken from Old Point Comfort between the York and James Rivers: provided, 1st, That the enemy's vessel *Merrimack* can be neutralized; 2d, That the means of transportation sufficient for an immediate transfer of the force to its new base can be ready at Washington and Alexandria to move down the Potomac; 3d, That a naval auxiliary force can be had to silence, or aid in silencing, the enemy's batteries on the York River; and 4th, That the force to be left to cover Washington shall be such as to give an entire feeling of security for its safety from menace. (Pollard 264; War Department, ser. 1, vol. 11, pt. 3, p. 58)

Lincoln was no more enthusiastic about the Peninsula Plan than about the Urbanna Plan. He never formally approved the Peninsula Plan, but, as Stanton put it, the president "makes no objection" to it.

Again, Lincoln gave McClellan additional instructions regarding the movement's execution. McClellan was to leave enough men in and around Washington to prevent the Confederates from threatening the city and to prevent them from recapturing Manassas. Remaining non-committal to the specific Peninsula Plan, Lincoln further directed McClellan to move "at once in pursuit of the enemy by some route" (Williams, *Lincoln* 73).

THE *MONITOR* AND THE *MERRIMACK* (*VIRGINIA*)

Of the conditions and reservations identified by the council of war and by the president, only the *Merrimack* issue seemed to be resolved without too much difficulty. The *Merrimack* was a 3,500-ton, forty-gun U.S. steam frigate that the Federals burned and scuttled when they abandoned Gosport Navy Yard on April 20, 1861. Confederate engineers had little trouble raising the hulk and found it to be in good shape except for the upper works, which had been destroyed by the fire. Confederate engineers raised the hulk and began its conversion to an ironclad (Quarstein 11–13).

Confederate Secretary of the Navy Stephen Mallory developed the idea to equip the *Merrimack* with armor and use it to break the ever-tightening Federal blockade. Naval Constructor John Luke Porter and Lieutenant John Mercer Brooke designed an ironclad ram that "made obsolete the navies of the world" (Boatner 560). Workers cut the hull down to the berth deck and built a casemate with slanting sides and ports for ten guns. The casemate walls contained twenty-four inches of oak and pine timber with four inches of armor plating. An open grating covered the top of the casemate to admit light and air to the gun deck.

Brooke armed the ironclad with two 6.4- and two 7-inch Brooke rifles and six 9-inch Dahlgren smoothbores. In addition, the casemate had a thirty-six-degree slope and was covered down to two feet below the waterline with overlapping plates of 2-inch armor. An armored pilot-house was forward, and protruding from the bow was a fearsome look-ing four-foot iron ram (Foote 255).

The major shortcomings of the *Merrimack*'s conversion were a draft of twenty-two feet and inadequate engines from the scuttled warship

(Wood 696). This and its great size severely limited its maneuverability, but such a beast could still wreak havoc with any Federal flotilla. To counter this threat, the Federal navy had proposed an attack on Norfolk to seize the Gosport Navy Yard before the Confederates could complete the *Merrimack* (Reed 102). The Federal army, however, did not support this operation, and the ironclad project continued without opposition. The preemptive opportunity was gone.

On March 8, the *Merrimack,* now rechristened the *Virginia,* sailed down the Elizabeth River into Hampton Roads on what was supposed to be a trial run. Its guns had not yet been fired, and workmen swarmed over its superstructure, busy with last-minute adjustments. It had a three-hundred-man crew, largely recruited from the army, and was under the command of Commodore Franklin Buchanan. At the age of sixty-two, Buchanan was known as the Father of Annapolis, because he had been instrumental in the founding of the U.S. Naval Academy and had served as the first superintendent there.

However, as the *Virginia* entered Hampton Roads, this "trial run" became something much more. Across the water Buchanan saw five Federal blockade ships lying at anchor. The *Minnesota,* the *Roanoke,* and the *St. Lawrence* lay off of Fort Monroe, and the *Congress* and the *Cumberland* lay off of Newport News. Buchanan was faced with an opportunity he could not resist.

First the *Virginia* went after the fifty-gun *Congress* and the thirty-gun *Cumberland,* making short work of both vessels. As the *Virginia* came within range, the *Congress* launched a well-aimed broadside that broke against the sloping armor with no effect. The *Virginia* continued its advance, impervious to the additional salvos from the *Congress* as well as the shots now coming from Brigadier General Joseph Mansfield's coastal batteries. When he was as close as he wanted, Buchanan opened the *Virginia*'s ports and menacingly exposed its guns. With these, Buchanan delivered a starboard broadside against the *Congress* and then rammed the *Cumberland,* leaving a hole that one officer described as large enough to accommodate "a horse and cart." When the *Virginia* swung clear, its iron ram beak broke off in the *Cumberland,* which began to fill with water. Called upon to surrender, her captain replied, "Never! I'll sink alongside." The *Cumberland* continued to fire as long as a gun remained above water but inevitably sunk, leaving its mainmast flag still flying defiantly above the water after the ship itself had struck bottom.

In the meantime, the wounded *Congress* had slipped its cable and had run aground trying to escape. The *Virginia*'s deeper draft forced it to remain at a two-hundred-yard distance, but the vessel nonetheless mercilessly raked the helpless *Congress* from end to end. With the captain dead, one of the *Congress*'s lieutenants ran up the flag of surrender. Mansfield's coastal batteries continued to fire even as the *Virginia* approached the surrendered *Congress* to take on prisoners. When one of Mansfield's lieutenants protested that since the *Congress* had struck its flag the Confederates had the right to take possession of it unmolested, Mansfield replied, "I know the damned ship has surrendered, but *we* haven't." Both Confederates and Federals were wounded by Mansfield's artillery, including both Buchanan on the *Virginia* and his brother, who had remained with the Union and was one of the officers on the *Congress*. In retaliation, the *Virginia* dropped back and set the wooden *Congress* on fire with red-hot cannonballs.

By now the other three frigates from Fort Monroe had entered the fray, but the *Virginia* caused them to run aground as they rushed to the battle. The tide, however, was beginning to ebb, and Executive Officer Lieutenant Catesby ap R. Jones broke off the *Virginia*'s attack against the *Minnesota* and withdrew toward the deeper waters of the Elizabeth River. The *Congress* burned into the night, creating an eerie glow, until the fire reached the magazine and the frigate exploded. The Federal squadron awaited the next day with apprehension, but by then the odds would be a little more equal (Foote 256–57).

Having learned of the Confederate efforts to build an ironclad, the Federals, led by Swedish-American inventor John Ericsson, began a similar project. Overcoming a three-month Confederate head start, Ericsson built the *Monitor* in less than a hundred days, just in time to check the *Virginia*'s rampage. The *Monitor* had a long, flat hull with no more than a foot or two of freeboard. In the midship section was a revolving iron turret mounting two eleven-inch guns. Aft of this was the smoke pipe and forward was a stubby iron pilothouse. The *Monitor*'s most significant advantage over the *Virginia* was a twelve-foot draft and high maneuverability. The *Monitor* had a sixty-man crew, all of whom had volunteered directly from the navy, in contrast to the *Virginia*'s largely ex-army crew. The *Monitor*'s captain was Lieutenant John Worden, a twenty-eight-year navy veteran recently released from seven months in a Confederate prison. Nine days after being commissioned, the *Monitor* was towed

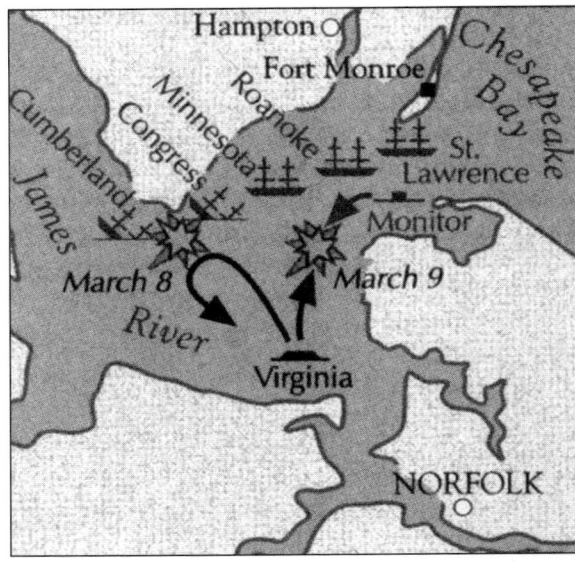

from New York to the Chesapeake by tug and steamed past the Virginia capes late on the afternoon of March 8 (Foote 259–60).

The ironclads presented an almost comical picture as they approached each other. Bruce Catton writes that the *Virginia* looked like "a barn gone adrift and submerged to the eves" (*Civil War* 77). The most common description of the *Monitor*'s appearance was "a tin can on a shingle" (Foote 259).

With Buchanan wounded, command of the *Virginia* fell to Lieutenant Jones. Taking full advantage of its greater maneuverability, the *Monitor* scored several hits on the *Virginia*. John Wood, one of the lieutenants serving with the *Virginia*, reported that "The *Monitor* was firing every seven or eight minutes, and nearly every shot struck" (Wood 701). This pounding cracked the *Virginia*'s railroad iron armor but failed to penetrate the two-foot pitch pine and oak backing. The two combatants continued to duel indecisively for two hours before both ships withdrew for what amounted to a half-hour intermission.

In the second two-hour engagement, the *Virginia* attempted to ram the *Monitor*, but with the loss of the *Virginia*'s ram beak from the previous day's fighting, this proved ineffective. Jones then tried to take advantage of the numerical advantage of his crew size and made several attempts to

The *Monitor*

board the *Monitor,* which repulsed all efforts. Finally, Jones brought the *Virginia* to within ten yards of the *Monitor* and struck the Confederate vessel's pilot house at point-blank range with a nine-inch shell. Stationed immediately behind the point of impact, Lieutenant Worden personally felt the full effect of this concussion and commanded his helmsman to sheer off. The *Virginia* too had wounds to lick, and with the ebb tide running, it withdrew across Hampton Roads to Norfolk. Tactically, the battle had been a draw, but strategically it must be counted as a Federal victory (Foote 261–62). Although the *Virginia* remained a threat, anxiety over the vessel's potential to single-handedly destroy the Federal fleet had been abated (Boatner 560–61; Catton, *Civil War* 77–80). At least one obstacle had been lessened along the way of McClellan's Peninsula Campaign.

But the *Virginia*'s mere presence still impacted on McClellan's plans by denying him the use of the James, which, along with the York, was his waterborne line of communication toward Richmond. Having failed to halt the *Virginia*'s construction, McClellan later recommended obstructing the Norfolk channel to bottle up the *Virginia*. The navy said that it would cooperate in such a venture if the army first disposed of the batteries guarding the channel. The army replied that it would do so if the navy, using the *Monitor*, first reduced the batteries. The navy answered that it could not afford to risk the *Monitor* in such an action while the *Virginia* remained a threat. There the matter rested, with the services unable to reach any consensus on cooperation (Sears, *McClellan* 173). The idea of the army and the navy acting in true concert rather than sequentially against the *Virginia* never seemed to receive any consideration. For the time being, McClellan would have to remove the James from his calculations, and genuine army and navy cooperation remained elusive.

McCLELLAN'S SUSPICIOUS ARITHMETIC

With the strategic threat from the *Virginia* thus mitigated, the real obstacle to McClellan's plan became Lincoln's concern for Washington. The corps commanders differed on the subject of how many troops were required to protect the city, with the minimum figure being forty thousand and the maximum being over fifty thousand (Williams, *Lincoln* 72–73). The first blow to McClellan's troop strength came when Lincoln bowed to political pressure and shifted Brigadier General Louis Blenker's ten-thousand-man division from the Army of the Potomac to Frémont's Mountain Department to be used in a far-fetched operation against eastern Tennessee. McClellan appeared to take this decision in stride, writing to Lincoln, "I cheerfully acquiesce in your decision without mental reservation." However, McClellan did mark this as a sign that the president was weak willed, and, whether from this cause or another, McClellan began to adopt a cavalier and poorly thought-out attitude toward the defense of Washington (Catton, *Army of the Potomac* 104; Sears, *Gates* 33; Williams, *Lincoln* 77).

McClellan must be held to blame for failing to comprehend Lincoln and Stanton's deep concern for the safety of the capital. McClellan made no effort to meet with them to discuss the issue and explain his ideas (Sears, *Gates* 33). Instead, he waited until he was aboard ship on April 1 and ready to depart for the Peninsula before he furnished Lincoln with any accounting of the forces left to defend Washington. At that point, McClellan finally dispatched a letter that used some misrepresentative mathematics to list more than 73,000 men he felt were dedicated to complying with the resolution reached at Fairfax Court House and Lincoln's instructions. The 73,000 men McClellan computed consisted of 10,859 at Manassas, 7,780 at Warrenton, 35,467 in the Shenandoah Valley, 1,350 on the lower Potomac, and 22,000 in and around Washington (Williams, *Lincoln* 79).

The forces at Manassas and Warrenton were designed to satisfy Lincoln's injunction that the Confederates not be able to recapture Manassas. However in late March, Jackson's presence in the Valley had caused McClellan to send Major General Nathaniel Banks and a part of the Manassas-Warrenton contingent to meet this new threat. At the time McClellan wrote his letter, Manassas was almost devoid of Federal soldiers. To correct this deficiency, McClellan ordered 4,000 men from Washington to move to Manassas. This left 18,000 in Washington rather than the 22,000 he told Lincoln. To reach the 10,859 figure in Manassas, McClellan advised that 6,000 more be brought in from Maryland and Pennsylvania. The 4,000 lost from Washington could be replaced by a like number detached from New York. Thus, at Manassas, McClellan was counting men who were not really there, and in Washington he was counting men he had only recommended be sent there.

The numbers game continued with the 7,780 at Warrenton. McClellan counted those twice, once as part of the Manassas total and then again as part of Banks's Shenandoah Valley force. As for Blenker, McClellan included his division in with Banks's numbers. It was correct that Blenker was still in the Valley and could be detained by Banks if necessary, but the truth of the matter was that he was on his way to Frémont and was not actually a part of the Washington defense force.

As T. Harry Williams concludes, McClellan was playing "fast and loose and careless with all his figures." In reality, not counting Blenker or the Warrenton group twice, McClellan left about 50,000 men to defend Washington and its approaches. While this number is consistent with the upper figure decided upon by the corps commander at Fairfax Court House, only 29,000 of these were either in the city of Washington itself or at Manassas or Warrenton. Lincoln undoubtedly assumed that the figure arrived at by the corps commanders referred to men in the Washington forts and just south of the Potomac, not as far away as the Shenandoah Valley (Williams, *Lincoln* 79–80). McClellan never took the time to explain to Lincoln that a location could be defended from the location itself, by denying the enemy the approaches to the location, or by threatening the enemy elsewhere to prevent them from attacking the location. Understood in these terms, even McClellan's Peninsula Campaign could be explained in the context of defending Washington. This failure to address Lincoln's concern for Washington would come back to haunt McClellan as the operation progressed.

THE NAVAL SUPPORT

The other loose end from the Fairfax Court House resolution that was tragically never tied up was the requirement for naval support for the assault. McClellan's plan was to advance up the Peninsula to establish an advanced base at West Point, about fifty miles northwest of Fort Monroe. There the Pamunkey and Mattaponey Rivers met to form the York. Before reaching West Point however, McClellan would have to contend with the Confederate defenses at Yorktown. Yorktown was the key, and it would have to be taken at once.

To do so, McClellan determined that full naval cooperation was "an absolute necessity." He wanted the navy to concentrate "its whole available force, its most powerful vessels, against Yorktown. There is the most important point—there the knot to be cut." If the navy cooperated, Yorktown could be taken in a matter of hours. If not, it would have to be reduced by the lengthy siege process (Sears, *McClellan* 173).

What McClellan envisioned was something similar to what the Federal forces had recently done at Fort Henry, Tennessee, on February 6. Without any direct or formal command link, Brigadier General Ulysses Grant and Flag Officer Andrew Foote worked together to arrange all aspects of the attack and quickly captured the fort (Donovan 43).

During the actual attack, the navy provided the lion's share of the effort. With three unarmed gunboats and four ironclad river gunboats, Foote delivered a short bombardment that compelled the Confederates to surrender. Grant's force was delayed by muddy roads but was not even needed for the assault (Boatner 394). All that was required of the army was to mop up a few scattered pockets of resistance and then occupy the works (Catton, *Terrible Swift Sword* 273). However, on the Peninsula, McClellan and Goldsborough would be unable to replicate the cooperation that made Foote and Grant successful.

McClellan's idea for joint action was a good one; unfortunately, it advanced nowhere beyond the idea stage. As T. Harry Williams puts it, "After McClellan made his request for naval support, a lot of talking took place about enlisting it, but everybody seems to have misunderstood everybody else" (*Lincoln* 76). The blame for the breakdown has to be shared by all parties—McClellan, the Lincoln administration, and a command system incapable of managing a complicated joint service operation.

The sequence of events ran something like this. McClellan made his request for naval support, which prompted Lincoln to come to Alexandria to talk with the general (Hassler, *McClellan* 75). Likewise, Stanton sent agents to meet with the navy. Flag Officer Louis Goldsborough, commanding at Hampton Roads, promised his cooperation with army operations but understood that his primary mission was to keep the *Virginia* away from the transports and the landing area around Fort Monroe. McClellan took no steps to finalize the coordination and left Alexandria convinced that Goldsborough was prepared to shell Yorktown into submission (War Department, ser. 1, vol. 11, pt. 3, p. 9; Williams, *Lincoln* 76; Sears, *Gates* 31; Catton, *Terrible Swift Sword* 273). The general would be disappointed.

Determining the exact blame for the failed cooperation is a difficult undertaking. Rowena Reed has made perhaps the most exhaustive attempt, concluding that there is "a strong suspicion" that the navy promised some cooperation that it did not deliver. She makes this assertion based on three factors. The first is an April 7, 1862, letter from Brigadier General Keyes to his friend, Senator Ira Harris, in which Keyes complains bitterly about the navy's performance and states that he approved of the Peninsula Campaign at the March 13 council of war only after personally requesting and receiving the navy's promise to shell the Confederates out of Yorktown if necessary. Indeed, the third precondition of the agreement by McClellan's corps commanders was "that a naval auxiliary force can be had to silence, or aid in silencing, the enemy's batteries on the York River" (Reed 125–30).

Reed's second point is that Assistant Secretary of the Navy Gustavus Fox, under examination by the Joint Committee on the Conduct of the War in 1863, reported that he had no knowledge of McClellan's plans when in fact he had been party to at least the earlier Urbanna Plan discussions. This point, combined with Reed's third factor, which is her accusation that numerous key papers appear to be missing from the navy's official records on the subject, tends to suggest some degree of a cover-up (Reed 125–30). None of Reed's factors is singularly decisive, and the second two are rather speculative, but her research does highlight the confusion of the situation.

More important is Reed's conclusion that from the general tone of all his communications with the navy, "it must have been apparent to McClellan that Goldsborough did not want to use his powerful warships on the York." Indeed, McClellan appeared to have realized this, as evidenced by his March 22 instruction to his chief of artillery, Brigadier General William Barry, that "it is possible we cannot count upon the Navy to reduce Yorktown by their independent efforts, [and] we must therefore be prepared to do it by our own means" (Reed 128). To the student of joint operations, this is a truly revealing statement. In spite of the Fairfax Court House Resolution's suggestion that one course of action might be for the navy to "aid in silencing" the York River batteries—implying a cooperative effort—McClellan discusses only a navy action or an army action. He does not at all seem to consider the two services jointly attacking the Yorktown problem. This is a profound statement on the status of joint cooperation at this point in the Civil War.

In the final analysis, the blame for the failure of joint army-navy operations in the opening days of the Peninsula Campaign must rest with McClellan. He obviously had no reason to believe that the navy would support him as he had originally envisioned. Thus, his task became either to convince the navy to cooperate or to develop a course of action that did not depend on the navy. He really did neither. Instead, he merely hedged his bet, hoping for full navy cooperation but doing nothing to bring it about and at the same time readying himself for a siege operation that would forfeit the temporary advantage he gained from the amphibious movement. It was a nondecision. The result was an operation in which both the army and the navy participated but did not cooperate.

THE MOVEMENT

So, with the impending disasters of not complying with Lincoln's intent for the safety of Washington and not completing the coordination with the navy hanging over his head, McClellan began his amphibious movement. The move was a mixed bag. Although it was an impressive display of the North's strength, it was also time-consuming and at times chaotic (Catton, *Terrible Swift Sword* 263).

To review the time schedule, on February 3, McClellan sent his letter to Lincoln indicating his intentions to conduct an amphibious movement. On February 14, Stanton began advertising for the necessary vessels, but Lincoln did not issue the order to acquire them until thirteen days later (Nevins 48; Williams, *Lincoln* 65). This order was passed to Assistant Secretary of War John Tucker on the next day, and Tucker then began putting together the fleet (Nevins 48). It is another interesting footnote in the study of the Peninsula Campaign as a joint operation that this task fell to the secretary of war rather than the secretary of the navy.

McClellan reports that his original plan was "to commence the movement with [McDowell's] First Corps as a unit, to land north of Gloucester and move thence on West Point; or, should circumstances render it advisable, to land a little below Yorktown to turn the defenses between that place and Fort Monroe." However, because the "transports arrived very slowly, especially those for horses, and the great impatience of the Government grew apace, it became necessary to embark divisions as fast as vessels arrived, and I decided to land them at Fort Monroe, holding the First Corps to the last, still intending to move it in mass to turn Gloucester" (McClellan 168). After these delays and changes to the plan, the embarkation itself began on March 17, with McClellan personally departing on April 1 (Williams, *Lincoln* 78). McClellan reached Fort Monroe on the next day, and by April 4 he felt he had enough forces on the ground to begin his advance on Yorktown. The transfer was fully effected on April 5 (Nevins 48).

The total tonnage moved was impressive. When McClellan arrived on April 2, some fifty-eight thousand men comprising five infantry divisions and some cavalry, as well as about a hundred guns, were already disembarked and ready to move (Catton, *Army of the Potomac* 107). A total of 121,500 men would eventually make the move (Sears, *McClellan* 168). Along with these came nearly fifteen thousand horses and mules, more than eleven hundred wagons, forty-four batteries, rolls of telegraph wire, timbers for pontoons, medicine, and countless other supplies (Nevins 48). The only losses were eight mules that drowned when a barge floundered.

To accomplish this feat, Tucker had assembled a picturesque flotilla of more than four hundred transports, including ocean liners, bay and harbor steamers, tugs, barges, and schooners of almost every size, shape, and description (Catton, *Terrible Swift Sword* 263–64). Such a colossal move was without precedent, leading one British observer to liken it to "the stride of a giant" (Sears, *Gates* 24).

FORT MONROE

What made such a movement possible was Federal control of Fort Monroe. The fort itself was massive, and its strategic location off Hampton Roads gave the Federals a foothold on the Peninsula. With Fort Monroe, McClellan had a protected landing site and a stable base of operations from which to launch his campaign.

Collection of materials for what was originally known as Fortress Monroe had begun in 1818. Actual construction began in March 1819, during the presidency of Virginian James Monroe, for whom the fort is named. It was built following the general plans for the fortifications designed by Marshal Sebastien Vauban at Toul, France. The fort consisted of seven fronts and covered approximately sixty-three acres. The original armament was planned to be 380 guns, but this was later extended to 412 guns, which were, however, never all mounted. The fort housed a peacetime garrison of 600 men, with a planned wartime strength of 2,625. No fort in the United States could boast such size, and no fort in Europe not enclosing a town was larger. It was known as the Gibraltar of the Chesapeake.

Fortress Monroe received its first official U.S. Army garrison on July 25, 1823, when Company G, Third U.S. Artillery, was transferred from Fort Nelson near Norfolk to guard military convicts involved in construction programs. By 1825, Fortress Monroe's garrison was the largest in the United States, containing one-third of the artillery troops and approximately one-tenth of the entire U.S. Army. In 1832, the secretary of war gave Fort Monroe its official designation, stating that "the work at Old Point Comfort be called Fort Monroe and not Fortress Monroe." Around this same time, from 1831 to 1834, Robert E. Lee was assigned there as a lieutenant of engineers assisting in the construction of the fortifications.

At the time hostilities began at Fort Sumter, South Carolina, the Fort Monroe garrison numbered four hundred men. Fort Monroe's size and land connection made it more easily defendable than Fort Sumter, and indeed it proved too powerful for Virginia's militia to overcome. The Virginians seemed to accept this reality, striking an informal agreement with Colonel Justin Dimick, Fort Monroe's commander, by which if Dimick did not encroach on Virginia soil, they would not deny him access to his water supply (West 80). Thus, as the Civil War erupted, the strategic fort remained in Federal hands. In the words of Clifford Dowdey, "Virginia had an enemy implanted on her soil" (*Land* 100). Similarly, John Quarstein and Dennis Mroczkowski describe Federal possession of Fort Monroe as the "key to the South" (27).

In May 1861, General Scott ordered Major General Benjamin Butler to reinforce Fort Monroe. Butler arrived on May 22, and by June 6 he had 6,750 soldiers under his command. Approximately 1,400 of Butler's force were located at Fort Monroe, while the remainder were located two miles to the east at Camp Hamilton, established at present-day Phoebus, to accommodate the overflow. A similar camp was later established at Newport News on the western corner of the Peninsula and named Camp Butler after the Federal commander (Roberts 816–17). From these positions, the Federals controlled the entire northern shore of Hampton Roads and the entrance to the James. To sustain this control, the War Board initiated a series of measures to ensure that Fort Monroe was provisioned with six months of supplies and otherwise improved to withstand a siege (Hattaway and Jones 130–34).

On June 8, 1861, Butler dispatched a strong reconnoitering party of forty-four hundred men toward Colonel John Magruder's defenses around Yorktown. The Federal troops encountered Colonel D. H. Hill's First North Carolina Infantry and Major George Randolph's Richmond Howitzers at Big Bethel. The Federals were repulsed in this first battle since Fort Sumter. Magruder, who was promoted to brigadier general, and his men were catapulted to instant fame. After this engagement, the Federal force at Fort Monroe remained in place until the initiation of McClellan's Peninsula Campaign the next spring (Long 146). In so doing, the Federals voluntarily denied themselves what could have been an excellent reconnaissance force and intelligence-gathering capability.

The Federals did have another intelligence-gathering source at Fort Monroe. Accompanying Butler was Thaddeus Lowe's ballooning rival,

John La Mountain, who provided the army with its first effective balloon observation on July 31, 1861 (Davis, *Civil War* 51–52). La Mountain is also credited with using the first "aircraft carrier" when he hitched his balloon to the *Fanny,* an armed transport, and rose above the waters of the Chesapeake on August 1, 1861. He also conducted the first nighttime aerial reconnaissance, this time anchored to the tug *Adriatic* near Fort Monroe.

La Mountain, however, soon ran into difficulties of both a physical and professional nature. While experimenting with hydrogen to try to develop a means of still higher ascensions, La Mountain suffered serious burns, and officials challenged the accuracy of his reports. His service to the army ended in early 1862 (Halsey 80).

The Big Bethel debacle gave General Scott, who had a long history of frustration with Butler, the needed excuse to move Butler to a less visible position. Scott decided to replace Butler with the superannuated Major General John Wool. Wool was junior in rank only to Scott and had to be called out of retirement to take command at Fort Monroe. Wool arrived at Fort Monroe on August 18, 1861, and would be in command there for the initiation of McClellan's Peninsula Campaign (War Department, ser. 1, vol. 11, pt. 3, pp. 18, 24, 65; West 104). The reorganization of the Federal army on March 11 established Wool as an independent commander. Furthermore, instructions issued on April 3 as the result of fears for the defense of Washington made Wool's forces, which had by that time climbed to ten thousand, unavailable to McClellan.

On June 6, 1862, Wool was replaced by Major General John Dix, and Dix's men were placed under McClellan's command. This was undoubtedly a pleasant development for McClellan, not just because it increased his troop strength but because Wool had made a habit of challenging McClellan's exaggerated estimates of Confederate troop strength (Sears, *McClellan* 197).

THE TERRAIN

The object of McClellan's move was an area rich in history. The peninsula between the York and the James Rivers had given birth to America's first plantation grandees, and the rivers continued to define the region. They would be fundamental to the campaign. If used correctly, these rivers could give McClellan an advantage of excellent water communications that Scott had lacked at Vera Cruz. Control of these rivers could allow McClellan to turn the Confederate positions by waterborne movement and then keep his troops resupplied by the rivers as well (Hattaway and Jones 145). Indeed, the rivers were the key, so much so that Freeman concludes, "the defense of [the Peninsula] was simple, so long as the James and the York were closed to the enemy" (Freeman, *Lee's Lieutenants* 1:695).

The York was the northern of the two main rivers. Just before its mouth, the river narrows, and there Gloucester lies on the northern bank and Yorktown on the southern. Batteries at these two locations, hardly a thousand yards apart, could control the river. At the beginning of the York is the town of West Point. From West Point, the Richmond and York River Railroad runs twenty-three miles west to Richmond. These features made the York attractive to McClellan as a line of supply.

The James River lay to the south. Hampton Roads, the area between Fort Monroe and Norfolk, is three miles wide and is formed by the confluence of the James River flowing southeast from Richmond, the Nansemond running up from the southwest, and the Elizabeth from the south. On the west side of the Elizabeth lay the key Gosport Navy Yard of Portsmouth. The section of Virginia south of the James is largely associated with Norfolk in the same way the north side is linked with Richmond. About eighty miles lay between Fort Monroe and Richmond. If Federal gunboats could push up the James, they would be in a position to both shell Richmond and support a land advance.

Movement by land, however, would not be easy. The peninsula between the York and the James was a fairly flat land traversed by creeks and small rivers. By the time of the Civil War, the great plantations were gone, tobacco having depleted most of the soil. The people had turned to general farming, and Richmond had become more and more important to their lives. Still, the area was sparsely populated, with farms few and far between and almost three-quarters of the land in timber.

The roads in the Peninsula had seen little in the way of development. The one that ran up its center (and later became Route 60 stretching from Williamsburg to Richmond) largely followed an Indian footpath and was in fact called the Pocahontas Trail. Closer to the river had sprung up a network of roads that converged on Richmond. These had been built based on the whims of farmers and were a haphazard collection that appeared suddenly and disappeared without warning (Dowdey, *Land* 166–67).

Those familiar with the area knew that these were not the types of roads that could support serious offensive operations, especially this early in the year. General Lee, serving in the capacity of military adviser to President Davis, wrote on March 18 that "the roads are hardly firm enough yet to invite [McClellan's] advance by land" (Dowdey and Manarin 132). Lee knew whereof he spoke. He had made an overland trip from Arlington to Fort Monroe in 1834 and reported himself "up to my ears in mud" (Freeman, *R. E. Lee* 1:120). McClellan, however, did not have such an appreciation of the local conditions.

Instead, McClellan based his terrain analysis on a faulty map prepared by one of Wool's topographical engineers, Colonel Thomas Jefferson Cram. This map depicted the Warwick River, a significant obstacle that lay perpendicular to much of McClellan's path, as being innocuously parallel to the James (Nevins 57–58). Also not depicted were the areas in which the Warwick and its tributaries had turned the marshy, heavily wooded flatlands into a quagmire (War Department, ser. 1, vol. 11, pt. 3, pp. 22–24). Furthermore, the Warwick had been dammed at different points, causing flooding in the intervening lowlands as far as Lee's Mill, where the river spread into marshlands. These dams were easily defensible points from which to hinder any attacker's advance (Longstreet 67).

West of the Warwick lay Williamsburg and after that another river, the Chickahominy. Even more so than on the Warwick, the banks of the Chickahominy were home to numerous swamps, small creeks, marshes, sharp ravines, and tangled growths of honeysuckle and underbrush that greatly restricted movement.

As for the "roads," a steady downpour quickly turned them into canals, useless for the passage of military traffic (Dowdey, *Seven Days* 45). McClellan found nothing of the sandy—and therefore passable in wet weather—highways he had been told to expect (Catton, *Army of the Potomac* 99). Artillery carriages and wagons soon became mired up to the axles, and one officer reported seeing a mule sink so deep in what was supposed to be a main road that only his ears were visible (Catton, *Army of the Potomac* 107).

The military aspects of terrain are commonly analyzed by considering the factors of OCOKA: observation and fields of fire, cover and concealment, obstacles, key terrain, and avenues of approach (FM

7-8 2-8). The terrain on the Peninsula can be analyzed using the same methodology.

Confederate observation and fields of fire would be excellent. They would be able to observe the Federal amphibious movement and know where it landed. This would rob the Federals of a certain amount of strategic surprise. Moreover, the bluffs along the river provided excellent fields of fire. The woods adjacent to the roads in the Peninsula would provide the Confederates with excellent cover and concealment. An even larger problem for the Federals would be the numerous obstacles in their path, primarily the Warwick and Chickahominy Rivers and the adjacent swamps and thick undergrowth.

Key terrain at the strategic level would be Richmond, which, as the Confederate capital, represented the campaign's objective. At the operational level, Fort Monroe provided the Federals with a foothold on the Peninsula, and Norfolk provided a home port for the *Virginia*. At the tactical level, key terrain included the high ground at Yorktown and Gloucester that provided positions for batteries to control the York River and the dam sites and bridges that allowed passage of the smaller rivers.

The avenues of approach were both on land and water. Those on land were limited. They would not withstand heavy military traffic, especially in rainy weather, and did not provide direct routes to Richmond. The water avenues of the York and the James were much more promising. However, high bluffs at Yorktown and Gloucester on the York and Drewry's Bluff on the James offered defensible terrain.

Weighing all these factors leads to the conclusion that the terrain on the Peninsula would favor both the defender and a smaller army contending against a large one (Stiles 26). This disadvantage would severely hamper McClellan's operation, but, even more significantly, his number juggling had already come back to haunt him.

CHAPTER 3

THE PLAN BEGINS TO UNRAVEL

On April 1, the day after McClellan left Alexandria, Brigadier General James Wadsworth, commander of Washington's defenses, reported to Secretary Stanton that the force to defend the city was inadequate in size and quality. He specifically complained about having to send four thousand men to Manassas. Stanton forwarded the report to Major General Ethan Allen Hitchcock, chairman of the Army Board, and Brigadier General Lorenzo Thomas, the adjutant general, and asked them if McClellan had complied with Lincoln's instructions regarding the safety of Washington. Hitchcock and Thomas replied that McClellan had not and stated that the capital was in fact in danger.

On April 3, Lincoln conferred with Stanton and the Army Board. The president directed Stanton to detain one of the Army of the Potomac's two corps awaiting embarkation at Alexandria and incorporate it into Washington's defenses. Stanton issued orders for McDowell's thirty-thousand-man corps to stay in place (Williams, *Lincoln* 81–82). That evening, McClellan received more bad news. The ten-thousand-man garrison force at Fort Monroe, previously promised to him, was no longer available for his use (Bailey 93).

With these twin losses, McClellan would start up the Peninsula with only a 90,000-man force rather than the 130,000 he had originally expected (Catton, *Civil War* 64). His complaints fell on deaf ears. However, McClellan must accept a large portion of the blame for his loss. In spite of his knowledge of Lincoln's misgivings about the Peninsula Plan and fears for Washington, McClellan refused to take the president into his confidence and give him a sense of security (Williams, *Lincoln* 81–83).

Had McClellan taken the time to explain the situation to Lincoln, the general could have made a strong case that Washington was in good hands. The only threat to Washington that could possibly materialize was Jackson in the Shenandoah Valley, and Banks should have been able to meet such a contingency. As such, Banks was an integral part of the Washington defenses, but McClellan never explained this to Lincoln. Likewise, the forces at Warrenton and Manassas served as a protective screen for Washington. Again, McClellan never explained the matter to Lincoln in these terms (Williams, *Lincoln* 81). There was much more to the defense of Washington than just the forces garrisoning the city and its surrounding forts.

Moreover, McClellan's offensive would contribute to the defense of Washington. McClellan had the advantage of water protecting his flanks and a secure supply line (Catton, *Army of the Potomac* 99). Therefore, his operation would be a penetration, capable of sustaining itself for a great length of time. Johnston, however, could conduct no more than a raid against Washington. Even if he captured the city, his stay would have to be temporary because his line of communication would be severed (Hattaway and Jones 88).

Lincoln too must accept some of the blame. If he was going to let McClellan carry out his plan, the president should have given the general the freedom to do so. If Lincoln had as serious doubts as his constant interference would indicate, he would have been better off replacing McClellan with a general in whom the president had more confidence (Williams, *Lincoln* 74).

ALLAN PINKERTON'S FAILED INTELLIGENCE

One of McClellan's most fateful decisions was to select Allan Pinkerton as his chief intelligence officer, the head of the Army Secret Service. Pinkerton ran the famous detective agency in Chicago, and he and McClellan were acquainted before the war through their mutual work with the Illinois Central Railroad. Unfortunately for McClellan, "Pinkerton was a fine man for running down train robbers and absconding bank cashiers but was completely miscast as chief of military intelligence" (Catton, *Army of the Potomac* 120–21). As a result, Bruce Catton concludes, Pinkerton "ran perhaps the most unaccountably inefficient intelligence service an American army ever had" (*Hallowed Ground* 138).

Pinkerton was behind from the beginning. When the Urbanna Plan gave way to the plan for a landing at Fort Monroe, the intelligence focus had to shift. The lower Peninsula had previously not been much of a priority, and Pinkerton found only three reports on the area in his files. Conspicuously absent was any militarily significant information about the terrain and the Confederate defenses along the Warwick (Sears, *Gates* 29). One asset that could have been used to correct this intelligence gap was the Federal force garrisoned at Fort Monroe. However, since Big Bethel, this force had remained close to the fort, thus denying the Federals a valuable reconnaissance opportunity.

A more enduring problem was that Pinkerton's intelligence system relied almost exclusively on a vast network of spies. In this regard, he gained impressive access to the Confederacy, with one of his men even winning the confidence of Confederate Secretary of War Judah Benjamin (Catton, *Army of the Potomac* 121). Unfortunately, this single-source-type intelligence lacked the coordination and corroboration essential in gaining an accurate intelligence picture. The fault, however, is not all Pinkerton's. McClellan raised no questions and seems to have been completely satisfied that he knew what he needed to know (Sears, *Gates* 29). Peter Parish notes that "Pinkerton made it his business in peace and war to provide his clients with the kind of information they wanted," and McClellan wanted to portray a large Confederate force (192). T. Harry Williams adds, "This inability to see things as they were is the key to the whole McClellan problem. He saw everything as he wanted it to be. Almost literally he lived in a world of make-believe" (Murfin 48). Parish continues that in McClellan's mind, "his skill in combating these reputedly vast enemy hordes elevated him in his own estimation and in the eyes of his army. Inflated estimates of enemy strength excused his delays and glorified his successes. They also seemed to demonstrate the folly of the administration in failing to support him fully" (192–93).

Only later, when Major General Joseph Hooker took command of the Army of the Potomac and appointed Colonel George Sharpe to succeed Pinkerton, did the Federal army begin to benefit from coordinated intelligence from prisoners, spies, deserters, contrabands, refugees, infantry and cavalry reconnaissance, signal stations, the balloon corps, and other

Allan Pinkerton

sources. Hooker would eventually benefit from a true picture of the Confederate army that McClellan lacked (Sears, *Chancellorsville* 69–70).

In contrast to McClellan, Lee made excellent use of accurate intelligence. Douglas Southall Freeman considers intelligence to have been "the beginning of all of [Lee's] strategic combinations" (Smith 139). Indeed, Lee had some personal experience with reconnaissance. As a captain in the Mexican War, he had found a rugged road that allowed the Americans to turn the Mexicans' left flank at Cerro Gordo. Of Lee's exploits, General Winfield Scott wrote, "I am impelled to make special mention of the services of Captain R. E. Lee, engineers. This officer, greatly distinguished at the siege of Vera Cruz, was again indefatigable, during these operations, in reconnaissance as daring as laborious, and of the utmost value" (Freeman, *R. E. Lee* 1:247–48). In the Civil War, not only did Lee appreciate the value of intelligence, he knew that his was better than McClellan's (Smith 155). This allowed Lee to get inside of McClellan's decision cycle and maintain the initiative by making decisions more rapidly than McClellan could.

Jeb Stuart's cavalry was Lee's most important intelligence source (Sears, *Gates* 167). McClellan too had plenty of cavalry, but it was all in fragments, with no centralized organization, focus, or direction. First, McClellan declared his volunteer cavalry fit only for picket and courier duty. Then he retained three squadrons of his regular cavalry for his personal escort and assigned the entire Second U.S. Cavalry to the provost marshal. He split his remaining three cavalry units among two brigades. To make matters worse, McClellan created a confusing command structure in which George Stoneman was his chief of cavalry but had only administrative duties (O'Neill, 49–50). Under such conditions, McClellan's cavalry was not much of a combat multiplier.

Stuart, conversely, truly was Lee's eyes and ears. On the Peninsula, Stuart proved his worth by riding all the way around McClellan's army and informing Lee that the Federal right wing lay exposed to the north and west (Catton, *Hallowed Ground* 138–39). Before dispatching Stuart on this mission, Lee told Stuart to "remember that one of the chief objects of your expedition is to gain intelligence for the guidance of future operations" (Freeman, *R. E. Lee* 1:247–48). Likewise, Stuart's reports of masses of supplies being burned at White House indicated to Lee that McClellan was changing his base (Freeman, *Lee's Lieutenants* 1:633; Eckenrode and Conrad 71). Freeman notes that Stuart consistently provided Lee with explicit answers to his questions, "soundly buttressed by meticulous reconnaissance and by intelligent observation" (*Lee's Lieutenants* 1:635). When it came to gathering timely and accurate intelligence, McClellan simply had no such capability as Stuart provided Lee.

One intelligence area that plagued both sides and probably Lee in particular was the lack of accurate maps. As previously mentioned, McClellan's map showed the Warwick as flowing parallel rather than perpendicular to his path. The map also did not show the areas where the Warwick and its tributaries had turned the surrounding area into a quagmire (Nevins 57–58; Dowdey, *Seven Days* 44). In Bruce Catton's words, when McClellan arrived at Fort Monroe, "The first thing he discovered was that someone had steered him wrong about those sandy roads on the peninsula" (*Army of the Potomac* 107). His entire picture of the terrain was wrong. Thus, in a letter to Stanton, McClellan complained that because of his lack of proper maps, he had been "obliged to grope" his way up the Peninsula (Hattaway and Jones 194; War Department, ser. 1, vol. 11, pt. 3, p. 151). Warren Goss, one of McClellan's privates, supports the commander's assessment, writing that in early April, "the topography of the country ahead was but little understood, and had to be learned by reconnaissance in force" (Goss 193).

The poor maps probably just reinforced McClellan's preconceived notion to go slow. For Lee, who was actively seeking offensive opportunities, the map problem was more important. Freeman observes that the "absence of reliable maps proved . . . serious throughout the campaign," citing examples of the problem's impact at Mechanicsville, Glendale, and Gaines' Mill as well as its personal effect on Magruder, Jackson, and other commanders (*R. E. Lee* 2:233). Archer Jones, perhaps overgenerously, attributes part of Longstreet's failure at Seven Pines to his being "without proper maps, a situation usual throughout the war" (63). D. H. Hill sums up the Confederate situation on the Peninsula, saying, "Throughout this campaign we attacked just when and where the enemy wished us to attack. This was owing to our ignorance of the country and lack of reconnaissance of successive battlefields." Lee wrote that "prominent among [the reasons for McClellan's escape] is the want of correct and timely information" (Freeman, *R. E. Lee* 2:232). Clifford Dowdey blames the Confederates' lack of accurate maps on Johnston's failure to map the region during his occupation of it. The result he says was that "the invaders . . . possessed more accurate maps than the natives" (*Land* 193). The poor maps would make reconnaissance and other forms of intelligence all the more important for both sides.

The great strategist Sun Tzu's famous military maxim, paraphrased as "Know your enemy" (82) is applicable to the Peninsula Campaign. The small prewar officer corps gave Federal and Confederate commanders intimate knowledge of their opponents. Longstreet, for example, reported to President Davis, "I knew General McClellan; that he was a military engineer, and would move his army by careful measurement and preparation; that he would not be ready to advance before the 1st of May" (Longstreet 66). Lee likewise understood McClellan. Stephen Sears notes that "Lee had taken careful measure of the general he faced. He had seen the excessive caution that marked all of McClellan's movements on the Peninsula. He noted his preoccupation with siege warfare and his dependence on the Richmond & York River Railroad to transport his supplies and heavy guns. He knew from Northern newspapers passed through the lines the inflated estimates of Confederate numbers that

Potomac army headquarters had given correspondents. In his offensive he would make his target McClellan himself as much as McClellan's army" (*McClellan* 207).

McClellan, however, lacked this ability to know his enemy. For example, on May 27 he assured Stanton that Johnston would not attack, explaining that, "I think he is too able for that." Just four days later, on May 31, Johnston surprised McClellan with an attack that began the battle of Seven Pines (Sears, *McClellan* 193). Lee, McClellan concluded, was "too cautious & weak" when faced with responsibility (Sears, *Gates* 57). He could not have been more wrong. McClellan did not know his enemy.

In the final analysis, it was as if "the Confederates knew McClellan as well, if not better, than did Washington" (Murfin 50). McClellan was nothing if not predictable, and the essence of military intelligence is predicting what your enemy will do. McClellan made this very easy for the Confederates. Moreover, he could not reciprocate this skill. The Confederate intelligence advantage would be a decisive factor on the Peninsula.

THE DEFENDERS

The full impact of McClellan's intelligence failure would be felt later in the campaign. What immediately mattered to McClellan when he disembarked at Fort Monroe was John Magruder's Army of the Peninsula. Reports of Magruder's strength vary slightly. Magruder's reports of 11,000 or 11,500 failed to include the reinforcement of Wilcox's brigade that had been provided him by Johnston. Freeman puts Magruder's strength at 12,000 (*R. E. Lee* 2:10), and Sears figures it to be 13,600. What is significant is that McClellan had received a report from Fitz John Porter on March 30 that the enemy possessed "no greater force this side than 15,000 men." This was a fairly accurate estimate. Sears notes that this "proved to be the only time during the entire Peninsula campaign that Federal intelligence came anywhere close to an accurate count of the opposing army. After this singular moment of reality, the Army of the Potomac would always confront a phantom Rebel army that existed only in the mind of the general commanding" (*Gates* 30).

Magruder's mission was twofold. He was to defend the country between the York and James Rivers against a Federal advance by land from

John Magruder's headquarters at Yorktown

Fort Monroe and to prevent Federal ships from moving up the York to West Point, where they could land troops less than forty miles from Richmond (Freeman, *R. E. Lee* 2:10). Thus, Magruder's plan involved three lines of defense. The first, located about twelve miles from Fort Monroe, was his lightly constructed "advanced line." It ran from Young's Mill on Deep Creek to Ship's Point on the Poquoson River. Federal Private Goss disdainfully questions whether these "insignificant rifle-pits" could even be called "fortifications" (190), and Sears calls the line "more sham than substance." In any case, it would require at least twice the force available to adequately man it, and its primary value was to keep the Federals at Fort Monroe at arm's length and unable to observe the main defensive line Magruder was constructing further up the Peninsula (Sears, *Gates* 26).

Indeed, this second line was much more formidable. The Warwick-Yorktown line ran from Mulberry Island on the James and followed the Warwick River to within a half mile of Yorktown. In all, the line covered fourteen miles. Yorktown was fortified with a series of redoubts, some of which were built on top of British works remaining from the 1781 siege

Confederate fortifications at Yorktown reinforced with cotton bales

during the American Revolution. Companion works were built across the York at Gloucester Point. Magruder had about half of his force positioned in the works at these locations. This left a mere five thousand or six thousand men stretched along the ten-mile front south of Yorktown (Dowdey, *Seven Days* 42). Fort Crafford on Mulberry Island and the Skiffes Creek redoubts defended against a flank attack on the James, and the Yorktown and Gloucester Point batteries protected Magruder's York River flank. However, Magruder lacked adequate heavy artillery and enough troops to prevent a combined attack by the Federal land and naval forces (Foote 399).

Key to Magruder's defense in this region was the river the Federals had so misjudged, the Warwick. The Warwick originally was a small and sluggish river, but two local farmers had built gristmill dams across it at

Lee's Mill and Wynne's Mill, and the Confederates had recently constructed three more. These served to widen and deepen the river, making it easily crossable only at the dams. At these key locations, Magruder had positioned artillery and infantry dug in behind earthworks (Bailey 94). Furthermore, the Confederates sank barges at the mouth of the Warwick to block Federal landing vessels (Foote 399). Such obstacles would make it slow going for the Federals.

In case his lower defenses were overcome, Magruder constructed a third defensive line about two miles east of Williamsburg. While allowing that "Magruder was doing the best that he could," Shelby Foote views these defenses rather incredulously, saying, "in fact it was not so much a line as it was a sort of rally-point in case the first gave way" (399–400). The entire line consisted of a series of fourteen redoubts, complete with supporting redans and rifle pits. The center was anchored by Fort Magruder (Redoubt #6), astride the Williamsburg Road. There, College Creek on the right flows into the James and Queen's Creek on the left flows into the York. Along this line, the Peninsula is eight miles wide, but the ground near the York especially is very broken, so the line was more defensible than might otherwise be imagined (Eckenrode and Conrad 33). The exceptions were at the locations of mill dams, which were passable by wagons. The redoubts on the left of Fort Magruder controlled the dam in Queen's Creek at Sander's Pond, but the dam in College Creek lay uncovered (Longstreet 68).

Forty miles behind Magruder's Williamsburg line, Robert E. Lee had another line under construction, this one stronger than any of Magruder's and immune from naval attack. It was anchored on its right by the James and on its left by the Chickahominy. A boggy stream also covered a portion of the front with a tributary known as White Oak Swamp. It was a good position, but it was also just ten miles east of Richmond. Lee did not want to use this position unless he had to and then only after the defenses further down had bought him time to gather reinforcements (Foote 400).

The resulting defense then was what is known doctrinally as a delay. A delay is "a form of retrograde in which a force under pressure trades space for time by slowing the enemy's momentum and inflicting maximum damage on the enemy without, in principle, becoming decisively engaged" (FM 3-0 8-7). Lee planned to use the delay as an economy of force measure until he could determine the enemy's main effort, build

White Oak Swamp

up his forces, and shape the battlefield for his offensive. These purposes are completely consistent with current doctrine (FM 3-0 8-7).

The defense was well tied into the terrain, and existing obstacles had been reinforced and covered by fire. The relatively small number of troops was used to cover likely avenues of approach, leaving more restrictive terrain uncovered. The positioning of forces provided interlocking fields of fire along these avenues of approach. In all, it was a fairly well laid out defense that took as much advantage of the terrain as possible to mitigate a paucity of troop strength.

Magruder, however, naturally was quite cognizant of being outnumbered. As soon as he began receiving reports of one troop-laden transport after another off-loading troops and equipment at Fort Monroe, he wired Lee that Yorktown must be abandoned at once (Dowdey, *Land* 166). Lee had only assumed his duties as Davis's military adviser on March 13, and now he already was faced with a crisis (Dowdey and Manarin 127).

On March 18, Lee reassured Magruder that "Notwithstanding the demonstrations of the enemy in your front, I see nothing to prove that he intends immediately to attempt your line. He is feeling your strength and desires to prevent your occupying other points. If strong enough, his feint may be converted into a real attack. I hope you will maneuver as to deceive and thwart him." To these calming words, Lee added a reminder about the poor conditions of the roads (Dowdey and Manarin 132). Lee knew his enemy.

On March 24, Lee received a telegram from Major General Benjamin Huger, commander of the thirteen-thousand-man Department of Norfolk, containing the information that he had seen more than twenty steamers advance down the Chesapeake Bay the previous evening and disembark troops near Fort Monroe. Huger's mission was to defend the city of Norfolk and the Gosport Navy Yard there. He had reason to fear the recent developments. Ambrose Burnside's success at Roanoke Island in January had left Huger vulnerable to a Federal attack up the inland waterways from Albemarle Sound, North Carolina (Freeman, *R. E. Lee* 2:10; Foote 225). In addition to this report from Huger to Lee, Magruder had also sent word to the secretary of war that he now believed the force opposing him had climbed to thirty-five thousand (Freeman, *R. E. Lee* 2:13).

The Federals were obviously up to something, but with such sketchy information Lee was not sure exactly what. He would have to consider three possibilities:

1. McClellan might have detached troops to cooperate with Burnside in North Carolina.
2. The new troops might have no connection with Burnside's movements and might be designed to join forces with the ten thousand already garrisoned at Fortress Monroe in an attack on Norfolk or up the Peninsula while McClellan advanced on Richmond from the north.
3. The reinforcements now facing Magruder might be the advance guard of McClellan's entire army, which was preparing to attack up the Peninsula (Freeman, *R. E. Lee* 2:13).

Within thirty-six hours of receiving news of the landing, Lee developed and initiated a plan to reconcentrate the Confederate forces to meet the still-undetermined Federal threat. On March 25, he sent warning orders to both Johnston and Huger advising Johnston to "organize a part of your troops to hold your present line, & to prepare the remainder to move to [Richmond], to be thrown on the point attacked" and telling Huger to "prepare to receive [the troops concentrating in Richmond] and determine the points to which they will be directed" (Dowdey and Manarin 135). Lee made it clear to both generals that the ultimate Federal objective was unknown. In fact, as late as April 1, he told Major General Theophilus Holmes, commander of the Department of North Carolina, that he believed the Federals' "real object is to attack Norfolk from both sides" (Dowdey and Manarin 141). Yet in spite of this degree of uncertainty, Lee took the critical step of getting the wheels of an organized plan in motion.

Lee decided that

1. Holmes would be strengthened to block Burnside's advance into North Carolina and to prevent his cooperation with any actions in Virginia.
2. Both Magruder and Huger should be alerted to prepare to reinforce each other should the Peninsula or Norfolk prove to be the Federal objective.
3. As soon as the *Virginia* returned from dry dock, the vessel would cover the mouth of the James to prevent the Federals from interfering with any of Magruder's or Huger's troop movements across the river. In the meantime, Lee sought to improve the James River batteries, accumulated transportation assets, and selected a troop crossing point that he hoped was above the reach of the Federal gunboats.
4. What few reserves were available would be sent to Magruder.
5. Magruder should be instructed to defend as far from Richmond as possible on a line running along the Warwick River to Yorktown. This line would not be voluntarily abandoned unless turned by Federal gunboats via the York or James. If this occurred, Magruder would withdraw to a line along the Chickahominy River. Along the way he would destroy the river landings and position his artillery on the riverbanks to halt any Federal advance up the James or the Pamunkey.
6. Most significantly, the bulk of Johnston's army would move from the Rapidan to the lower Peninsula and attack the Federals there. If, however, enough of McClellan's army remained in northern Virginia to threaten Richmond from that direction, Johnston would be prepared to return his troops to their original position (Freeman, *R. E. Lee* 2:5–16; Hattaway and Jones 161).

This reconcentration would eventually bring the Confederate strength opposing McClellan to fifty-three thousand (Freeman, *R. E. Lee* 2:23).

However, when McClellan disembarked on April 2, he had fifty-eight thousand men ready to move at Fort Monroe, compared to Magruder's eleven thousand (Nevins 56). When McClellan began his march to Yorktown on April 4, he had about seventy thousand men, although Clifford Dowdey believes that barely sixty thousand "were ready to move" (Williams, *Lincoln* 90; Dowdey, *Seven Days* 43). McClellan reports that he had "53,000 men in condition to move" (169). Even allowing for McClellan's lower figure, at this juncture the amphibious move clearly presented the Federals with a very advantageous situation.

YORKTOWN

McClellan's original intention was immediately to take Yorktown with the help of the navy (Williams, *Lincoln* 90). When he arrived at Fort Monroe, he met with Goldsborough to finalize the plan. By McClellan's account, Goldsborough said that he could neither protect the James as a line of supply nor provide vessels to help reduce the batteries at Yorktown and Gloucester by either bombardment or threatening their rear (Hassler, *McClellan* 86). McClellan reports that Goldsborough "could only aid in the final attack after our land batteries had essentially silenced [the Confederates'] fire" (McClellan 169).

Part of the reason for Goldsborough's conclusion was his belief that his main effort was to protect the Federal force from the *Virginia,* which left him with only seven wooden gunboats to support other army operations. These were adequate for furnishing escort and fire support for landing but were by no means up to challenging shore batteries. In fact, their guns could not even elevate sufficiently to reach the batteries on Yorktown's bluffs (Sears, *Gates* 32).

Bruce Catton's research yields a different account of the meeting between McClellan and Goldsborough. According to Catton, instead of discussing a bombardment of Yorktown, McClellan asked for the navy's help in reducing the Confederate fort at Gloucester. Between Gloucester and Yorktown, the York is a mere one thousand yards wide, so the twin Confederate forts effectively sealed the mouth of the river (*Terrible Swift Sword* 273).

By Catton's account, McClellan briefed Goldsborough that the general wanted to land troops on the banks of the Severn River, a few miles north of Gloucester, and then assault the city from its rear. Success here would ideally lead to the surrender of Yorktown but at the very least would allow gunboats to advance up the York.

Goldsborough assigned the seven wooden gunboats for the Severn River operation, and on April 4 McClellan told McDowell, who still had not left Alexandria, that his corps would be used to attack Gloucester. Not until after issuing this order did McClellan learn that Stanton had retained McDowell's corps for the defense of Washington (Catton, *Terrible Swift Sword* 275). There would be no First Corps attack on Gloucester.

Thus, the final operation had nothing of the joint army-navy flair originally envisioned. Whether you prefer McClellan's recollection that Goldsborough could not help or Catton's conclusion that McClellan changed his plan and then felt stymied by the loss of McDowell, the result was the same. There was no naval bombardment, and there was no Severn River operation. What actually happened was a very uncomplicated maneuver in which one column marched up the Peninsula on the right toward Yorktown, while another marched on the left toward Williamsburg (Nevins 56). This left column was led by Erasmus Keyes, commander of the Fourth Corps. The aim was to turn Magruder's flank and push on to Halfway House between Yorktown and Williamsburg. When Keyes gained control of the road network around the Halfway House area, the Federals hoped that Magruder would be forced to withdraw from Yorktown (Dowdey, *Seven Days* 43–44).

In addition to having to deal with the loss of McDowell, a torrential rain, the disappointing roads, and the faulty maps, McClellan had no prize in Keyes, whom Clifford Dowdey describes as "an uninspired career soldier" (*Seven Days* 45). Keyes's march toward Halfway House could also be described as "uninspired."

Late in the afternoon of April 5, Keyes's advance element came under fire from artillery and entrenched infantry at Lee's Mill, where the maps indicated only a harmless depot. At dusk, Keyes's force collapsed in an unorganized halt among the wooded swamps east of the Warwick. Based on this contact, Keyes reported, "Magruder is in a strongly fortified position behind Warwick River, the fords to which had been destroyed by dams, and the approaches to which are through dense forests, swamps, and marshes. No part of his line as far as discovered can be taken by assault without an enormous waste of life" (Dowdey, *Seven Days* 45).

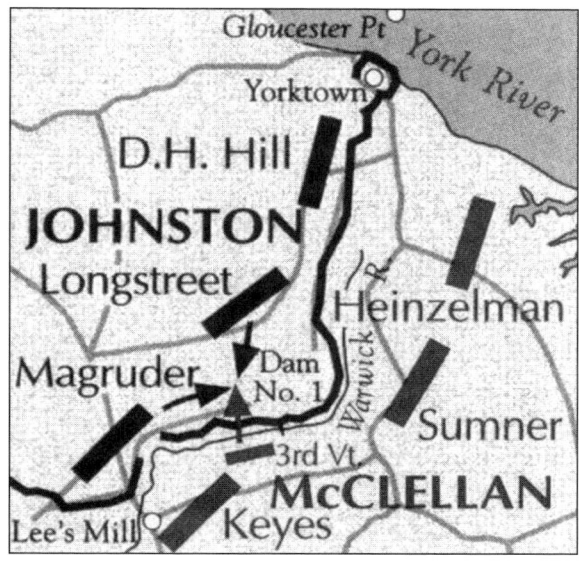

Keyes's word choice is important. T. Harry Williams opines that McClellan "loved [his men] so much that he did not want to hurt them. [This quality] made him forget that soldiers exist to fight and possibly to die. The trouble with McClellan was that he liked to think of war as bloodless strategy" (Murfin 52).

Of course such a brief contest could not have possibly warranted such a dramatic conclusion on Keyes's part. Had Keyes and the other Federal commanders probed a little more aggressively, they would have found that Magruder was in fact stretched quite thin—so thin, in fact, that he had to resort to a ruse just to keep up appearances (Williams, *Lincoln* 90).

While Magruder was "never especially distinguished as a combat general," he did have quite a reputation as an amateur thespian (Catton, *Army of the Potomac* 108). Perhaps remembering Lee's advice to "maneuver as to deceive and thwart" the enemy, Magruder put his acting talent to work (Dowdey and Manarin 132). On April 5, he marched a couple of his regiments out of a thicket and into a clearing within view of the Federal advanced guard. The Confederates then disappeared into another woods, doubled back around while still out of sight, and then repeated the process (Catton, *Army of the Potomac* 108). For such antics, Dowdey dubs Magruder the "Great Demonstrator," and Sears writes that

"No one could match [Magruder] at stretching limited resources into plausible illusions" (Dowdey, *Land* 193; Sears, *Gates* 25).

Enemy commanders and the decisions they are expected to make will always be a target for military deception (FM 3-0 11-17), and it is always easier to trick commanders into believing something they were predisposed to believe (Joint Pub 3-58 IV-2). B. H. Liddell Hart notes that the "real target is the mind of the enemy commander, not the bodies of his troops" (Joint Pub 3-13 II-4), and the Federal commanders, infected by Pinkerton's exaggerated estimates of Confederate strength and McClellan's predilection for defense, were readily persuaded. Bruce Catton notes that "McClellan believed that he was outnumbered because it was his nature to overestimate his disadvantages. As a result, it was impossible for him to take advantage of his opportunities" (*Hallowed Ground* 138). It was an environment perfect for deception, and Magruder was just the man to pull it off.

The primary victim of this deception was Third Corps commander Samuel Heintzelman over on the right, but, downplaying this event, Dowdey reserves most of his criticism for Keyes. Dowdey goes so far as to say that had Keyes "been placed in that position by Stanton solely for the purpose of drenching McClellan's offensive flare, he could not have been more effective" (*Seven Days* 45). Lincoln's corps reorganization, which filled the command positions with officers unsympathetic toward McClellan, certainly had to have a negative effect on the aggressive execution of the campaign.

In a major instance of failed intelligence, McClellan derived from his commanders' reports that Johnston had arrived to reinforce Magruder. Such was not the case, for Johnston's advance division did not join the defense until April 10 and other divisions did not arrive for several weeks thereafter. For the first six or seven days of the operation, McClellan enjoyed a four-to-one numerical superiority over the Confederates (Nevins 58; Dowdey, *Seven Days* 46). In spite of this advantage, late on April 5 McClellan elected to begin preparations for a siege, even writing to his wife to request that she send him his books on Sevastopol (Sears, *Gates* 48). The British observer who had described the Army of the Potomac's first move in the campaign as "the stride of a giant" would pronounce the second step as "that of a dwarf" (Sears, *Gates* 39). After the war, McClellan explained his decision, writing that he was "obliged to resort to siege operations in order to silence the enemy's artillery fire, and open the way to an assault" (171). If this in fact was McClellan's critical concern, a closer examination of this Confederate artillery threat is in order.

McClellan appears to have overrated the Confederate artillery. Along the Warwick-Yorktown line, Magruder had eighty-five pieces of heavy artillery and fifty-five field guns (Sears, *Gates* 26). Douglas Southall Freeman calls Magruder's collective artillery "antiquated smoothbore guns" that could be outranged and destroyed by the "superior Federal ordnance" (*Lee's Lieutenants* 1:148).

Even what little artillery the Confederates had was unreliable. In October 1861, Magruder wrote to Josiah Gorgas, chief of Confederate ordnance, that about half the artillery shells fired by his command exploded at the muzzle (Wiley 303). Magruder also complained of the instability of the gun carriages, noting that some of his cannon would "dismount themselves after a few fires" (Wiley 300).

For these and other reasons, Bruce Catton concludes that McClellan "could certainly have overwhelmed Magruder with one push, and even after Johnston got the rest of his army to the scene the Confederate works could probably have been stormed" (*Civil War* 64). But as Johnston reported on April 22, "No one but McClellan could have hesitated to attack" (Freeman, *Lee's Lieutenants* 1:154). One of McClellan's early biographers notes that "McClellan had now arrived at the most critical point of his career as a commander of an army in the presence of an enemy" (Murfin 49). McClellan would not rise to the occasion.

Instead of being genuinely concerned about the strength of Magruder's artillery, McClellan made the siege decision because he had favored that course all along. Keyes's and Heintzelman's underdeveloped reports merely gave McClellan the excuse he needed to put the plan in motion. Brigadier General W. F. Barry, McClellan's chief of artillery, remembered the siege decision as follows: "The army having arrived in front of the enemy's works went into camp, and preparations were at once commenced on the siege. From this date until April 10 active reconnaissance of the enemy's lines and works were pushed by the commanding general . . . for the purpose of selecting a suitable place for the landing of the siege train" (Dowdey, *Seven Days* 46). From the time of the initial contact with the enemy, McClellan obviously had very little interest in engaging in a battle of maneuver.

McClellan later sought to blame his stagnation on the loss of McDowell, arguing that the absence of this corps "left me incapable of continuing operations which had begun. It compelled the adoption of another, a different and less effective plan of campaign. It made rapid and brilliant operations impossible. It was a fatal error" (Catton, *Terrible Swift Sword* 275). Such a complaint smacks of after-the-fact rationalization (Dowdey, *Seven Days* 47). True, McClellan reports that his original plan was for McDowell to move from Alexandria first, but since March 17 McClellan had known that this movement plan had been changed (McClellan 168). If McClellan was really interested in fighting a campaign of maneuver, he had plenty of time to reassign McDowell's mission to another corps. Yet McClellan's inflexibility got the best of him. If McDowell was no longer available, so be it. No modifications would be made.

At McClellan's request, Lincoln did release one division, Franklin's twelve-thousand-man command. These troops took ten days to make the trip, and, upon arriving on April 20, they remained on transports for two weeks while McClellan, his staff, and the navy rehashed the Gloucester operation (Catton, *Terrible Swift Sword* 277). There was obviously something more to McClellan's cautious approach than the loss of McDowell.

This factor was McClellan's predilection for siege, with all its slow, deliberate execution and its small loss of life (Williams, *Lincoln* 90). These were the lessons McClellan had learned at Vera Cruz and Sevastopol. In fact, there is much to suggest that McClellan departed Washington with the idea of a siege already fixed in his mind (Nevins 58). Evidence of this intention can be found in the composition of McClellan's siege train, only half of which was regular siege guns. The other half comprised seacoast guns weighing up to eight tons, which could not be moved at all in the field and had to be fired from stationary platforms. Such monsters could not be employed in a battle of maneuver against Magruder's Warwick River line, but they were the very type of heavier guns McClellan saw that Scott originally lacked at Vera Cruz (Dowdey *Seven Days* 51). McClellan came ready for a siege.

The siege lasted until May 3, when, as at Manassas, the Confederates abandoned Yorktown and withdrew up the Peninsula on their own terms. By that time, Lee had affected his reconcentration. Johnston gives great credit to Magruder for this turn of events, writing that Magruder's "resolute and judicious course . . . saved Richmond, and gave the Confederate Government time to swell that officer's handful

Federal thirteen-inch seacoast mortars at Yorktown

to an army" (Stiles 25). In fact, the real credit for this opportunity lay with McClellan more than it did with Magruder. As a result of his over-caution, McClellan lost the tremendous advantage he had gained in early April from his amphibious move.

In addition to resorting to siege rather than maneuver warfare, there are two other reasons why McClellan forfeited this advantage. McClellan's almost complete failure to incorporate joint army-navy operations into any planning has already been discussed. The remaining flaw was his characteristic failure to plan for the unexpected. In this regard, McClellan had committed himself to a siege extremely early in the campaign. He had no plan for a pursuit should the enemy refuse to cooperate and not remain in place for the bombardment.

This lack of flexibility is a key flaw of McClellan's generalship. Most of the blame for this failure can be laid on McClellan's personality, but fault can also be found with the absence of any semblance of branches

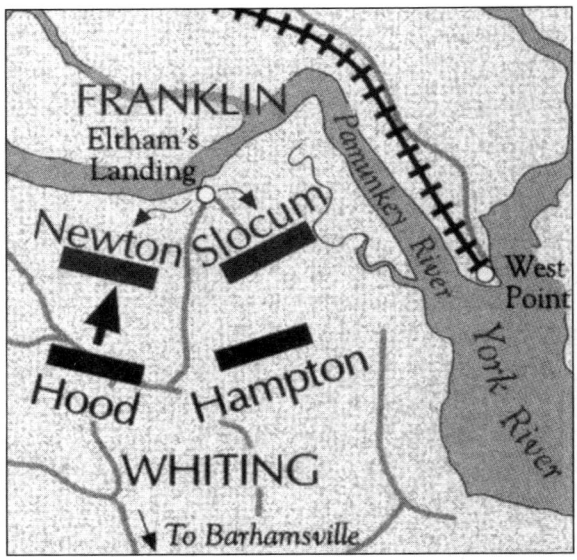

and sequels in his planning. Branches are options built into the basic plan. They add flexibility to plans by anticipating situations that could alter the basic plan as a result of enemy action, availability of friendly capabilities or resources, or even a change in weather. Related to the branch is the sequel. A sequel is a subsequent operation based on possible outcomes—victory, defeat, or stalemate—of the current operation (Joint Pub 3-0 III-20). A Confederate withdrawal from Yorktown represents one situation that McClellan could have developed as a branch to his plan.

A branch that would have addressed such a development would have been to move a force, likely Franklin's only recently disembarked division, by ship up the York to cut off the retreating enemy (Sears, *McClellan* 182). The reason such a movement had not previously been tried was the batteries at Yorktown and Gloucester. With these defenses gone, nothing stood in the way of a waterborne turning movement except McClellan's failure to seize the opportunity. McClellan's apologist, Warren Hassler, credits McClellan with "superintend[ing] the important movement of Franklin's division by water to West Point in an endeavor to cut off, if possible, the retreat of the enemy," but by the time McClellan put such a plan in motion, it was too late. Johnston had begun to withdraw on May 4, and Franklin's movement did not occur until May 7 (Hassler, *McClellan* 98).

Nonetheless, Hassler argues that "McClellan's stroke up the York River ensured [the Confederates'] rapid withdrawal northwestward toward Richmond" (Hassler, *McClellan* 104). This withdrawal is exactly what Johnston had intended, with or without Franklin's movement. Joseph Cullen correctly notes that Johnston "was only interested in the safety of his wagon train, and did not attempt to hold the terminus at West Point," where Franklin struck. Thus, Johnston attacked Franklin only because the Confederate general thought the Federals were attempting to cut off his withdrawal. Cullen notes that although Johnston's attack was repulsed, his withdrawal continued because Franklin's orders were to hold until reinforcements arrived (46). Had Franklin cut off the with-drawal, McClellan would have had something to be proud of, but instead, what became known as the battle of Eltham's Landing was nothing more than an inconclusive and belated sideshow rather than a preplanned and decisively executed branch of McClellan's original plan.

WILLIAMSBURG

With the decision to evacuate Yorktown, Johnston had the opportunity to pursue the strategy he had favored all along—to retreat rapidly to the immediate vicinity of Richmond and therefore negate the possibility of the Federals' outmaneuvering him by their command of the waterways and getting there first. Thus, Johnston felt a strong desire to put as much distance between the Confederate army and Yorktown as possible.

Johnston withdrew along two roads that came together eleven miles past Yorktown and two miles short of Williamsburg. For his retreat, he had his forces organized into four commands led by Longstreet, D. H. Hill, Gustavus Smith, and David Jones, who was replacing the ailing Magruder. Jeb Stuart's cavalry provided the rear guard (Sears, *Gates* 68).

It is usually hard for a withdrawing force, encumbered by all its logis-tical trains and equipment, to move as fast as a more lightly configured pursuing force. For this reason, the delay is often fought from a series of successive positions (FM 3-0 8-7). Williamsburg would become one such position for the Confederates.

Early on the afternoon of May 4, Federal troops pursuing Johnston caught up with Stuart's cavalry and began to push it in. Johnston realized that, especially given the hard rain, it would be a long time before his army could traverse the single muddy road leading from Williamsburg to Richmond. He would need to leave a force behind to check the Federal advance while the main body continued its retreat.

The nearest brigade that Johnston had available was that of Brigadier General Paul Semmes. Johnston led Semmes into Fort Magruder and likewise ordered Brigadier General Lafayette McLaws to bring up Brigadier General Joseph Kershaw's South Carolina Brigade and two batteries of artillery. This force was enough to temporarily check the Federal advance currently being spearheaded by Brigadier General William "Baldy" Smith. Darkness brought an end to the immediate Federal threat, but Johnston knew that the fight was far from over. He decided to reassign the rearguard mission to Longstreet and have Semmes and Kershaw's brigades continue the retreat. By 8:00 that night, Longstreet had brigades belonging to Brigadier Generals Richard Anderson and Roger Pryor in place, and Johnston's main body was back on the road (Harris 44).

Williamsburg was well suited for Longstreet's task. There the Peninsula was only seven miles wide. The area Longstreet was to defend was further narrowed by Queen's and College Creeks, leaving only three miles in between. These natural defensive conditions had all been improved by Magruder's earlier efforts. In all, Williamsburg was "a very ugly place to have to attack" (Sears, *Gates* 70).

The Federal command conducting the pursuit was confused. McClellan had elected to remain behind at Yorktown, belatedly planning Franklin's nonconsequential turning movement via the York (Sears, *McClellan* 181–83). Instead of taking command, McClellan put Edwin Sumner in charge of the pursuit. Sumner, however, had none of his Second Corps troops with him because McClellan was holding them for his planned York River operation. Instead, Sumner was commanding Heintzelman's Third Corps and Keyes's Fourth Corps, both of which were still being led by Heintzelman and Keyes.

The lead element of Keyes's corps was Baldy Smith's division. The other two divisions had been drastically slowed by the mud and were still well to the rear. Heintzelman was in a similar circumstance. Only Hooker's division was to the front. Philip Kearny's was some miles back, and his third division remained at Yorktown with McClellan. Thus, the Army of the Potomac would open the battle of Williamsburg with three corps commanders in command of two divisions totaling just 18,500 men. Whether the other three divisions could get up in time to influence the fight remained to be seen. Anyone would have been troubled by such a mess, but Sumner was overwhelmed. Stephen Sears concludes, "From first to last that day, General Sumner would be unable to grasp the battle unfolding in front of him" (*Gates* 71).

At 7:00 A.M. on May 5, Hooker moved forward on foot to observe the situation. With him was his artillery chief, Major Charles Wainwright. The only place the pair could find for the division artillery was on the Lee's Mill Road and in a nearby cornfield that faced the junction of the Lee's Mill and Yorktown Roads. As Wainwright moved his artillery into position, the Confederates opened fire.

Hooker brought forward more infantry and artillery and established fire superiority over the Confederates. He did not, however, storm Fort Magruder. Instead, he waited for Keyes's Fourth Corps to advance on his right to conduct a combined attack. This, however, would not occur.

The problem was that Lee's Mill Road and Yorktown Road did not come together at a narrow angle from which the two Federal columns could be within mutually supporting distance of each other. Instead, the roads turned sharply as they joined and met at almost a ninety-degree angle. Moreover, the junction was covered by Confederate fire. At their closest, Hooker and Smith were still a good mile apart and separated by swampy, restrictive woods. As Sears puts it, "They might as well have been on different battlefields" (*Gates* 70).

Another complication was that Hooker had somewhat jumped the gun and advanced ahead of the rest of the Federal troops. If Sumner indeed had a plan, it was for the divisions of Smith, Hooker, and Brigadier General Winfield Scott Hancock to lead the assault while the divisions of Brigadier Generals Philip Kearny, Silas Casey, and Darius Couch continued to move forward to provide reinforcements. For the time being, however, Hooker was alone (Harris 44–45).

With the Federals thus stymied, Longstreet seized the initiative. He already had two brigades at Fort Magruder, and he brought three more into action and one more into close reserve. Now, he put his senior brigadier, Richard Anderson, in immediate command of the offensive. Anderson sent Cadmus Wilcox's brigade on a flank attack through a sheltered ravine and into the thick woods off to the west. Captain John Pelham's Stuart Horse Artillery provided a devastating fire from the front as Wilcox crashed into Hooker from the west. Longstreet had brought to bear 10,350 Confederates against Hooker's 9,000. The Confederates had the momentum and the positional advantage. Hooker was in growing danger of having his flank turned (Sears, *Gates* 73).

Hooker called for reinforcements, but Sumner refused to part with any of the troops with him. Instead, any help would have to come from within the Third Corps, which meant Kearny's division. Kearny was marching as fast as he could, but the mud and overcrowded roads prevented him from getting into position until between 2:30 and 3:00 P.M. Led by five regiments of Brigadier Generals Hiram Berry's and David Birney's brigades, Kearny's division entered the woods and slowly forced the Confederates back. However, as the Federals emerged from the protection of the woods, they found themselves on an open killing ground. The Federal attack on the west was stopped. Shawn Harris sums up the fighting by saying that "all that Hooker and Kearny had accomplished was the needless slaughter of their own men for a piece of ground the Confederates could not have cared less about" (45–46).

But while this was going on in the west, developments were also occurring in the east. Earlier in the morning, a runaway slave had come into Sumner's headquarters and reported that a woods road crossed a dam on a branch of Queen's Creek called Cub Creek and went around the flank of the Confederate defenses. There was a Confederate redoubt built to cover the dam, but it was currently unoccupied. An engineer sent to confirm the report found it to be true. Anderson had either not seen this redoubt when he occupied his position or had assumed it was controlled by other troops. From it, the Federals could pour artillery fire into Fort Magruder and roll up the Confederates' eastern flank (Freeman, *Lee's Lieutenants* 1:179).

Smith urged Sumner to let him send a division to exploit this situation, but Sumner refused, feeling that such a move was too dangerous. Eventually, however, Sumner relented, allowing Smith to send a mere brigade. Smith made the most of the opportunity, sending Hancock, Smith's best brigadier, and quietly reinforcing his command to five regiments and two batteries.

Defending the Confederate eastern flank was Major General D. H. Hill's division. Longstreet had previously taken the precaution of ordering

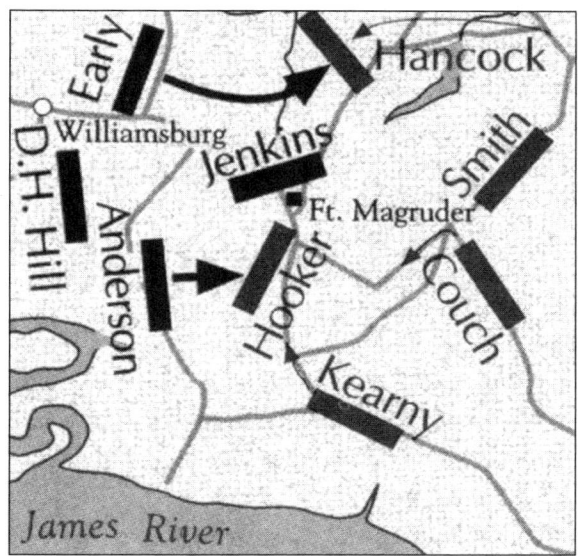

Hill to call up a reserve force, and Hill had done so in the form of Brigadier General Jubal Early's brigade. Thus, Early was now ready and waiting on the green of the College of William and Mary. As Hancock began to threaten his position, Hill called Early forward.

Hill and Early did not know the exact location of the Federal force, but they could hear its guns firing from behind a strip of woods. Early recommended a flank attack, which was approved by Hill, Longstreet, and Johnston, who had returned to the battle but was leaving its control in Longstreet's capable hands. At close to 5:00, Early's brigade began its march to the left and then turned into the woods in the direction of the sound of the guns. Early led the Twenty-fourth and Thirty-eighth Virginia regiments on the left, and Hill led the Twenty-third and Fifth North Carolina regiments on the right.

The rain and mist had already limited visibility, and darkness was now beginning to fall as well. Moreover, the woods were swampy and tangled with undergrowth. Control and orientation were difficult. Early emerged from the woods with the leftmost regiment, the Twenty-fourth Virginia, and found himself a quarter mile to the right of the Federals. Rather than flanking the Federals and attacking their exposed rear, the Confederates had ended up well in front of their enemy and were themselves exposed.

Without waiting for the rest of the brigade, Early ordered the Twenty-fourth Virginia to wheel to the left and attack into the Federal guns.

At first things looked good for the Confederates. They had caught a disgusted Hancock finally relenting in the face of Sumner's orders and pulling back to Cub Creek. Thus at this point, the situation looked to the Confederates like the Federals were retreating. Instead, Hancock took up a strong position and delivered a deadly fire on the Virginians. Early was hit in the shoulder and had to go to the rear.

When Hill and the Fifth North Carolina emerged from the woods, they fared no better. With a huge gap between himself and Early's Virginians and the Thirty-eighth Virginia and the Twenty-third North Carolina still in the woods, Hill ordered his men to wheel and attack. He was met by Hancock's thirty-four hundred rifles and eight artillery pieces. It was no contest, and Hill quickly broke off the attack. Before the Confederates could reach the safety of the woods, Hancock counterattacked. The Fifth North Carolina, being the most exposed, bore the brunt. It lost 302 of the total 508 Confederate casualties.

Jeffry Wert correctly questions the necessity of these losses, arguing that the Confederate mission at Williamsburg was to "secure the passage of the army's wagons by keeping McClellan at bay" (105). In fact, H. J. Eckenrode and Bryan Conrad report that when Hill first requested to send Early against Hancock, Longstreet demurred because doing so would be inconsistent with the assigned mission, and it would be a mistake to enter into a general engagement. Ultimately, however, Hill gained Longstreet's acquiescence (Freeman, *Lee's Lieutenants* 1:191; Eckenrode and Conrad 37). By the time of Early and Hill's attack, Wert notes that not only were the wagons well beyond Williamsburg but darkness was already bringing the day's action to a close. In fact, Douglas Southall Freeman concludes that when Early attacked, the Federals had already begun to withdraw (Freeman, *Lee's Lieutenants* 1:188). The impetuous Confederate attack was at this point unnecessary (Freeman, *Lee's Lieutenants* 1:188; Wert 105–6). Hancock's loss was only one hundred men, but fortunately for the Confederates, his repeated pleas for reinforcements had gone unanswered, and he attempted no pursuit (Sears, *Gates* 78–81).

By this time, McClellan had arrived at Sumner's headquarters, but the battle was now over. Having accomplished his mission of delaying the Federal pursuit, Longstreet quietly withdrew his force under the cover of darkness. As dawn broke on May 6, Federal pickets crept forward, only to find the Confederate defenses empty. Thereafter, Longstreet reported, "The pursuit was not active, hardly annoying." To assist in covering the Confederate withdrawal, Gabriel Rains had placed several primed shells hidden in abatis. In so doing, Rains initiated land mine warfare (Longstreet 79).

The battle of Williamsburg revealed several trends that would continue throughout the Peninsula Campaign. The most important of these was the need to achieve mass at the decisive point and time. Even though the Federals had the preponderance of forces, they were unable to bring their greater strength to bear on the battlefield because of terrain, weather, a lack of synchronization, and ineffective command. Thus, Longstreet was able to gain a 10,350-to-9,000 advantage over Hooker.

A second trend was the high cost of a frontal attack. This principle was generally understood at the time of the Civil War, and the turning movement, envelopment, and flank attack were considered to be the superior forms of maneuver. Throughout the Peninsula Campaign, the Confederates would endeavor to conduct these indirect forms of attacks, but, as Early experienced at Williamsburg, the fog of war would often intervene, and a frontal attack would result. The situation would be exacerbated when the attack was conducted piecemeal and the terrain supported the defense. The consequences were tremendously high casualty rates and usually failed attacks.

The third condition that showed itself at Williamsburg and would remain a theme throughout the campaign was McClellan's conspicuous absence from the immediate battlefield. Not only would McClellan not directly exercise command, but he would fail to establish an effective chain of command in his stead. This situation would become even worse as McClellan began his retreat after Mechanicsville.

The final factor of continued importance in evidence at Williamsburg was the terrain's inhospitality to the conduct of attacks. This situation was compounded by the lack of good maps and unfamiliarity with the confused road networks. The fact that Lee's Mill Road and Yorktown Road were effectively separated by terrain and distance right up to their final juncture was in large part the reason for the Federals' inability to achieve mass. The ineffectiveness of the Federal cavalry prevented the Federals from reconnoitering the battlefield and determining this situation in time to do something about it. Off the roads, the terrain was equally challenging, as Early and Hill learned in their counterattack against Hancock. Again, a lack of reconnaissance prevented the Confederates from knowing the exact location of the Federal force, and this uncertainty combined with the restrictive terrain to befuddle the attack. In terrain that already favored the defender, any imperfection in the attack would be magnified throughout the Peninsula Campaign.

Thus, the need to achieve mass at the decisive point, the high cost of frontal attacks, McClellan's abdication of direct command responsibilities, and the difficulty of attacking under the given terrain and conditions emerged as characteristics of the battle of Williamsburg. Freeman would call Williamsburg "a practice battle in which commanders learned what they should not do" (*Lee's Lieutenants* 1:191). Unfortunately, its characteristics would become trends throughout the Peninsula Campaign and would impact Confederates and Federals alike.

DREWRY'S BLUFF

In addition to McClellan's land force, the Confederates had to defend against a Federal navy advance on Richmond via the James River. This threat became a reality as Johnston withdrew up the Peninsula, abandoning Norfolk and leaving the *Virginia* without a home port. The vessel's draft of twenty-two feet was too deep for it to withdraw up the James, and finally, on May 11, the *Virginia* was abandoned and blown up. With its demise, the only remaining obstacle to a Federal advance up the James was Drewry's Bluff, sometimes called Fort Darling. The Confederates would turn Drewry's Bluff into a blocking position, "a defensive position so sited as to deny the enemy access to a given area or to prevent enemy advance in a given direction or an avenue of approach" (FM 101-5-1 1-20). The position was well suited for such a task.

Ninety feet high, Drewry's Bluff stood on the south bank of the James River less than eight miles south of Richmond. At this point, the river bends sharply to the east for a short distance and then turns again to the south. The property was owned by Captain Augustus Drewry, who would also command the Confederate gunners of the Southside Heavy Artillery during the battle.

The position began as a small earthen fort built by Chesterfield County farmers in 1861 in what Dowdey calls "an illustration of the individualism of the South working for it instead of, as usually, against it" (*Land* 182). By the time of the Peninsula Campaign, the fort was still only partly finished and lightly defended, but the Federal advance gave the workers a new sense of urgency. Lee's eldest son, Colonel George Washington Custis Lee, personally supervised the crews of soldiers, sailors, and laborers rushed into service.

The first step for the Confederates was to improve the natural obstacle created by the bend in the river. To do so, they sank several stone-laden hulks and drove piles to narrow the channel at critical points. Now, any Federal gunboats making the turn would have to expose their flanks to the fort.

The Confederates then placed guns from the scuttled *Virginia* and other weapons nearly one hundred feet above the water level, knowing that the Federal gunboats would be unable to elevate their guns high enough to hit the Confederate weapons. In all, the Confederates had four smoothbore and four rifled cannon trained on the river.

These defenses would be tested on May 15, when Commander John Rodgers and a Federal squadron of five vessels—the *Monitor* and *Galena* as well as the *Aroostook, Port Royal,* and *Naugatuck*—advanced on Drewry's Bluff. The *Monitor*, of course, was the star of the flotilla, but the *Galena* was also an ironclad. Designed by Samuel H. Pook for C. H. Bushnell and Company, the *Galena* had been commissioned on April 21. It was a corvette with unusual round sides and armor made of interlocking iron bars that reached a thickness of four inches on the sides. One observer described the vessel as looking "like a great fish with iron scales" (Guttman 3). The *Galena* had a two-mast schooner rig, two Ericsson vibrating-lever steam engines, and two boilers, generating eight hundred horsepower and driving a single screw to give a maximum speed of eight knots. The *Galena* was outfitted with two hundred-pound Parrot rifles and four nine-inch Dahlgren rifles. In spite of all this, Goldsborough was unimpressed, dubbing the *Galena* "a sad affair" (Guttman 3).

In addition to these two ironclads, Rodgers's flotilla had the three wooden vessels. The *Naugatuck* had originally served as the *E. A. Stevens* in the U.S. Revenue Cutter Service (the precursor of the Coast Guard) and had been built by John Stevens in 1844 as a single-screw ship, although it later received two screws. The *Naugatuck* weighed 192 tons, was 110 feet long, and had two inclined engines with one boiler. Its armament was one one-hundred-pound rifle, two ten-pound rifles, and

Confederate position at Drewry's Bluff

a twelve-pound howitzer. Its main protection was its ability to partially submerge by flooding compartments to increase its draft from seven feet, eight inches to nine feet, ten inches. Rounding out the fleet were the *Aroostook,* a screw gunboat, and the *Port Royal,* a side-wheel gunboat.

In a four-hour engagement, this picturesque Federal fleet proved no match for the defenders. The astute Confederate artillery placement carried the day. The *Galena* alone was hit forty-four times. This fire was especially effective because the Confederates delivered plunging fire down on the *Galena,* penetrating the ironclad's thin deck armor, while the Federal shells were unable to reach the Confederates high on the bluff.

However, this pounding did not deter Corporal John Mackie, a member of the Marine contingent aboard the *Galena.* Mackie delivered a steady stream of fire against the Confederates and became the first

Marine to win the Medal of Honor. His citation read in part, "As enemy shellfire raked the deck of his ship, Corporal Mackie fearlessly maintained his musket fire against the rifle pits along the shore and, when ordered to fill vacancies at guns caused by men wounded and killed in action, manned the weapon with skill and courage" (Guttman 1–7; B. Anderson 82–83; Dougherty, "Drewry's Bluff" 20–21).

Part of the Federal failure at Drewry's Bluff can be traced to the continued difficulty in establishing true coordination between the land and naval forces. The accurate fire of the Confederate heavy guns on the bluff and effective sharpshooting from the riverbanks proved too much for the unsupported naval attack. One Confederate officer, however, observed, "Had Commander Rogers [*sic*] been supported by a few brigades, landed at City Point or above on the south side, Richmond would have been evacuated" (*National Park Service*). For this reason, Freeman notes that Drewry's Bluff "showed the possibility of joint operations on James River by the Federal Army and Navy" and led General Lee to believe that the Federals might later initiate an operation there similar to Yorktown (*Lee's Lieutenants* 1:211).

The situation that could bring about Lee's fears would be if McClellan were to take advantage of the opportunity to shift his entire line of advance to the James. Lincoln, too, had seen this opportunity, later informing McClellan that he had hoped "that the opening of James River . . . with an open road to Richmond, or to you, had effected something in that direction." McClellan, however, had no such plans. Instead, he pushed up the York River to West Point and established a massive supply base there. This had been McClellan's plan from the beginning and probably represented somewhat of a holdover from his original Urbanna Plan.

Years after the war, McClellan seems to have finally understood the shortcomings of the York River line (Cullen 49–51). An attack based on the York would have to straddle the Chickahominy. Thus, from his base at West Point, McClellan was supplying four-fifths of his army across this obstacle. Conversely, with an attack based on the James, McClellan's left flank would have been protected by the Federal gunboats and his right flank, if he withdrew his whole force south of the Chickahominy and destroyed the bridges, would be guarded to a large extent by the river (Cullen 56–57). But with his typical inflexibility, McClellan would not adopt such a change to his original plan. Only years after the war McClellan began arguing that presidential interference had forced him to use the York River, split his army, and thus doom the campaign (Cullen 51).

McClellan seemed to have recognized the limitations of the York River line but stuck with it anyway. He wrote,

> The entire army could probably have been thrown across the Chickahominy immediately after our arrival, but this would have left no force on the left [north] bank to guard our communications, or to protect our right and rear. If the communication with our supply depot had been cut by the enemy, with our army concentrated on the right [south] bank of the Chickahominy, and the stage of the water as it was for many days after our arrival . . . the troops must have gone without rations, and the animals without forage; the army would have been paralyzed. It is true, I might have abandoned my communications, and pushed forward toward Richmond, trusting to the speedy defeat of the enemy and the consequent fall of the city, for a renewal of supplies; but the approaches were fortified, and the town itself surrounded with a strong line of entrenchments requiring a greater length of time to reduce than our troops could have dispensed with rations. (Stiles 29)

McClellan lacked the audacity required for such a course.

Whatever the line of advance, the Federal navy recognized that a joint force was required. Goldsborough opined, "Without the Army the Navy can make no real headway toward Richmond. This is as clear as the sun at noonday to the mind" (Sears, *Gates* 105). Indeed, the navy had requested "a cooperating land force" for the Drewry's Bluff operation, but McClellan had wired the War Department that he was "not yet ready to cooperate with them." This was in spite of his previous promise that the "Navy will receive prompt support wherever and whenever required." The Federal forces on the Peninsula were still a long way from working jointly.

Indeed, Joseph Cullen concludes that McClellan's refusal to cooperate in this case was all about who would get credit for the victory. This presents a marked contrast between Grant and Foote's selfless teamwork at Fort Henry. Cullen asserts, "At this point, it appears that McClellan was afraid the Navy would get credit for the capture of Richmond if the gunboats succeeded in reaching the city." In Cullen's view, by opting for a siege followed by a steady overland advance rather than providing a ground force to support a naval attack, McClellan was motivated by the idea that "All he needed was time and the fame and glory would be his" (47–48).

Jon Guttman describes a second factor contributing to the Federal failure as being "typically McClellanesque." This was McClellan's penchant for the slow, deliberate approach to any problem. He had instructed Goldsborough to have his men "reduce all the works of the enemy as they go along, spike all their guns, blow up their magazines," rather than bypass ineffective resistance in a swift and bold advance on Richmond. Thus, Rodgers devoted some attention to Fort Boykin, Fort Huger, Fort Powhatan, City Point, and Appomattox Manor as he closed on Drewry's Bluff. None of these were major efforts, but, as Guttman observes, "Each delay in the flotilla's progress brought that much more time for Richmond's mixed bag of troops and tars to shore up her defenses" (3). Perhaps more instructively for a student of McClellan's psyche, this methodical approach to warfare was born of the same philosophy that had likewise led McClellan to opt for a siege rather than a rapid advance on land.

CHAPTER 4

THE TIDE TURNS

If McClellan epitomized caution, Robert E. Lee epitomized audacity. Audacity is one of the army's four characteristics of offensive operations. It is "a simple plan of action, boldly executed [to] produce decisive results" (FM 3-0 7-6). When McClellan was faced with uncertainty, he became paralyzed. Audacious commanders "dispel uncertainty through action; they compensate for lack of information by seizing the initiative and pressing the fight" (FM 3-0 7-6). The effects of Lee's audacious generalship would soon bear fruit on the Peninsula.

THE IMPACT OF JACKSON'S VALLEY CAMPAIGN

Even from the beginning of McClellan's planning for an operation on the Peninsula, events in the Shenandoah Valley had been of great importance. The Valley represented a potential Confederate avenue of approach into Washington, and Lincoln had demanded that McClellan leave an adequate force behind to guarantee Washington's safety. Pursuant to this requirement, McClellan issued instructions to Major General Nathaniel Banks on March 16 ordering him to "open your communications with the valley of the Shenandoah. As soon as the Manassas Gap Railway is in running order, intrench a brigade of infantry, say four regiments, with two batteries, at or near the point where the railway crosses the Shenandoah. Something like two regiments of cavalry should be left in the vicinity to occupy Winchester and thoroughly scour the country south of the railway and up the Shenandoah Valley. The general object is to cover the line of the Potomac and Washington" (War Department, ser. 1, vol. 12, pt. 3, p. 859).

The defensive-minded Joe Johnston was certainly not contemplating an offensive against Washington. On March 1, Johnston, still unaware of McClellan's plans for an amphibious turning movement, expected the Federals to march directly south to Richmond. Accordingly, he ordered Jackson's Valley Army to fall back on line with the main army, protect its flank, secure the Blue Ridge passes, and slow or stop enemy progress up the Shenandoah. Of greatest importance was the fact that Johnston needed Jackson to prevent Banks from reinforcing McClellan. With Jackson commanding barely thirty-six hundred infantry in mid-March, Robert Tanner concludes that Johnston expected the Valley Army "only to be a sideshow in [Johnston's] retreat—if it was equal even to this minor task" (103). As it turned out, Jackson would become much more than a sideshow.

In fact, Jackson's Valley Campaign was nothing short of a masterpiece. For those interested in a detailed analysis of the campaign, James Robertson's *Stonewall Jackson: The Man, the Soldier, the Legend,* Frank Vandiver's *Mighty Stonewall,* and Robert Tanner's *Stonewall in the Valley* are excellent references. This study, however, is concerned with the impact of Jackson's actions on the Peninsula Campaign.

As President Davis's military adviser, Robert E. Lee saw an opportunity to use Jackson in the Valley to threaten McClellan's plans for the Peninsula. On April 21, Lee wrote to Jackson, "I have no doubt an attempt will be made to occupy Fredericksburg and use it as a base of operations against Richmond. Our present force there is very small. . . . If you can use General Ewell's division in an attack on General Banks and drive him back, it will prove a great relief to the pressure on Fredericksburg" (Deaderick 33; War Department, ser. 1, vol. 12, pt. 3, p. 859). This was such a momentous concept that Herman Hattaway and Archer Jones conclude that while "the Confederates were concentrating coastal troops at Richmond, their immediate response [to McClellan's Peninsula Campaign] began with Jackson's important Valley campaign" (176).

Taking full advantage of central position and interior lines, Jackson routed the Federal forces in the Valley to an extent that Lincoln feared for Washington. These events were occurring right as McClellan's efforts were beginning to show promise on the Peninsula, with the Confederates evacuating Yorktown on May 3, withdrawing toward Richmond, and in the process abandoning Norfolk. On May 18, McClellan had received a telegram from Secretary of War Stanton announcing that McDowell's

First Corps would be marching from Fredericksburg, where it had been held previously for fear of Washington's safety, and soon join him. But by May 24, as the result of Jackson's victories in the Valley, this had abruptly changed. Lincoln telegraphed to McClellan, "In consequence of Gen. Banks' critical position I have been compelled to suspend Gen. McDowell's movement to join you" (Sears, *Gates* 110).

McClellan complained that "the object of Jackson's movement was probably to prevent reinforcements being sent to me" rather than to attack Washington (Sears, *Gates* 111). McDowell agreed, stating, "It is impossible that Jackson can have been largely reinforced. He is merely creating a diversion and the surest way to bring him from the lower valley is for me to move rapidly on Richmond" (B. Alexander 73). Such arguments failed to convince Lincoln, who had never been comfortable with McClellan's provisions for Washington's safety, and Lincoln's order stood. Disgustedly, McDowell lamented, "If the enemy can succeed so readily in disconcerting all our plans by alarming us first at one point then at another, he will paralyze a larger force with a very small one" (Cullen 62). This is exactly what Jackson had done.

Douglas Southall Freeman sums up the importance of Jackson's impact by saying that Jackson "had used [his] small force so effectively that he had forced President Lincoln to change the entire plan for the capture of Richmond. At a time when the junction of McDowell with McClellan would have rendered the defense of the Confederate capital almost hopeless, Jackson temporarily paralyzed the advance of close to 40,000 Federal troops. Rarely in war had so few infantry achieved such dazzling strategic results" (*Lee's Lieutenants* 1:485).

SEVEN PINES

On May 28, Johnston held a meeting with his division commanders at his headquarters on Nine Mile Road about five miles east of Richmond. During the meeting, a courier arrived with news from Jeb Stuart that McDowell was not moving toward Richmond but was returning to Fredericksburg. This was the profound impact of Jackson's work in the Shenandoah Valley, which dramatically changed the strategic situation outside of Richmond. With McDowell's threat removed, the Confederates could act more aggressively.

On May 30, Johnston received word from D. H. Hill that his reconnaissance had revealed that the entire Federal Third and Fourth Corps were south of the Chickahominy but that no Federal units were located on the Charles City Road south of Seven Pines. Johnston concluded that the two Federal corps south of the river were vulnerable. This situation, coupled with the arrival of Huger's division from Petersburg, gave Johnston an opportunity to strike. If Federal reinforcements could be cut off across the Chickahominy, two-fifths of McClellan's army could be defeated in detail.

What made this an even greater possibility was that on May 30 a violent rainstorm struck the area, and the Chickahominy rose three to four feet above normal. The low, swampy bottomlands bordering the river became completely impassable. Corduroy roads disappeared in the mud, and swift currents ripped away bridges. Joseph Cullen notes, "It was a perfect time for an attack on the two corps south of the river, under the circumstances, it would be almost impossible to reinforce them" (54). Furthermore, McClellan appears to have been indisposed to move his Fifth and Sixth Corps from the north side of the Chickahominy, where they were extending to the Federal right, to meet McDowell if he were sent south, in spite of the fact that Lincoln had already told McClellan that McDowell would be retained (Murfin 51; O'Neill 47). McClellan had dangerously split his force, and weather and his own inflexibility were now making matters worse.

Johnston's attack plan would take advantage of three roads emanating from Richmond. The first, Nine Mile Road, was on Richmond's northeastern edge and passed through Fair Oaks Station on the Richmond and York River Railroad before intersecting with Williamsburg Road at Seven Pines. Nine Mile Road was named for the nine miles it ran from Richmond to Seven Pines. Fair Oaks was a mile northwest of Seven Pines and six miles from Richmond. To the south of Nine Mile Road was Williamsburg Road, which followed a generally straight course east from Richmond seven miles to Seven Pines and across the Chickahominy at Bottom's Bridge. At no time were Nine Mile Road and Williamsburg Road more than two and a quarter miles apart. The third road was Charles City Road, which branched off from Williamsburg Road roughly two miles beyond Richmond, traveling southeast. Between Charles City Road and Williamsburg Road, near Seven Pines, sprawled White Oak Swamp (Freeman, *Lee's Lieutenants* 1:225).

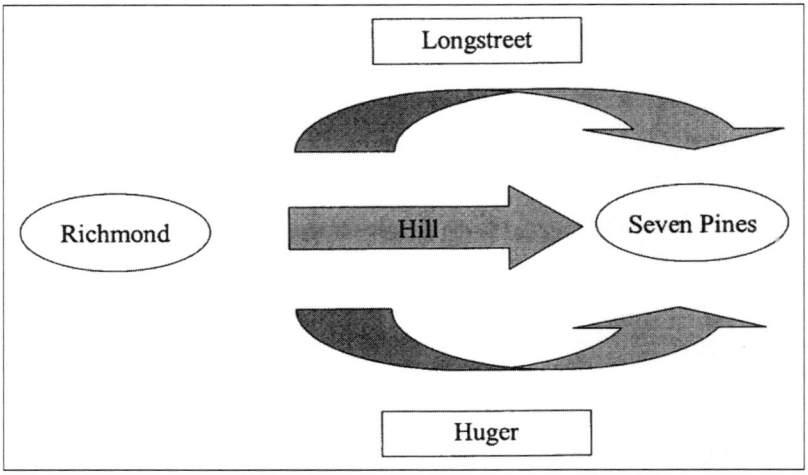

Joseph Johnston's planned double envelopment of Seven Pines

Johnston planned to use these roads to strike the enemy simultane-ously head-on and on both flanks (Sears, *Gates* 118). Such a maneuver is called a double envelopment. Hill's division would lead the assault down Williamsburg Road toward Seven Pines. Longstreet's division would advance on Hill's left wing to Fair Oaks using Nine Mile Road. Huger's division would move by Charles City Road on Hill's right. Longstreet commanded D. H. Hill's and Huger's divisions. Brigadier General Chase Whiting would lead Gustavus Smith's division in support of Longstreet, and Magruder would be in general reserve. Only A. P. Hill's newly organized division did not have an assigned role. If all worked according to plan, Johnston would be committing as many as 51,600 of his 74,000 troops against the two Federal corps numbering just under 33,000 south of the Chickahominy. The battle was to open at about 8:00 (Sears, *Gates* 119–20; Wert 110–13).

This was Johnston's first offensive of the war, and his plan appeared deceptively easy. Longstreet even commented that the "plan seemed so simple that it was thought impossible for any one to go dangerously wrong" (Longstreet 88). In fact, many things would go wrong, proving Clausewitz's assertion that "Everything in war is very simple, but the simplest thing is difficult" (Clausewitz 119). Sears notes, "Few battles ever go entirely as their generals plan them, but seldom does a battle stray as far from plan as Seven Pines on May 31, 1862" (*Gates* 138).

Johnston's challenge was one of synchronization and intent. In this regard, the Confederates would fail miserably. As Freeman frames it, the question was "would the commanders of the large, independent and poorly staffed Divisions be able to bring their troops together against the enemy?" (*Lee's Lieutenants* 1:229). The answer would be no.

Johnston's gravest error in this department was to give Longstreet only verbal instructions. Longstreet claims to have heard "every word and thought expressed," but events would indicate that a communication breakdown certainly occurred (Wert 113). Johnston did issue written orders to Hill, Huger, and Magruder, an interesting fact because Hill and Huger would be under Longstreet's command for the attack. Nonetheless, Johnston's orders to Huger did not even state the intention to fight a battle on May 31, leaving Huger greatly confused. Johnston also failed to clarify to Huger the command relationship, a serious omission and potential source of confusion given the fact that Huger outranked Longstreet. Sears observes, "Johnston was not a general noted for his attention to detail, and Seven Pines would demonstrate how careless he could be" (*Gates* 120).

Longstreet placed his troops in motion at first light on May 31, but instead of moving them east down Nine Mile Road, he sent them toward Richmond on a road that connected with Williamsburg Road. When Whiting's leading brigade reached Nine Mile Road, it encountered some of Longstreet's troops still breaking camp. Whiting had expected his path to be clear; unable to find Longstreet, he sent an aide to Johnston seeking clarification. Johnston sent back word that "Longstreet will precede you," but not until 10:00 did Johnston learn where Longstreet was. At that time, Johnston received a message that Longstreet and his brigades were on Williamsburg Road at the Charles City Road intersection. Exasperated, Johnston grumbled, "I wish all the troops were back in camp" (Wert 115).

In addition to befuddling Whiting, Longstreet's march had brought him into the path of Huger's division. Longstreet's men stopped to build a bridge across the rain-swollen Gillies Creek and then filed nearly fourteen thousand troops across one by one, slowing their advance long enough for Huger's division to bump into them. Huger halted his men and went to find Longstreet. He located him at Hill's headquarters, where Longstreet told Huger that Longstreet was moving to attack with Hill and that Huger should follow to Charles City Road and support the main assault. This was the first Huger knew of Longstreet's role in the attack. Johnston's orders to Huger had left him in the dark. Moreover, Longstreet's instructions to Huger would effectively remove a division of between five thousand and seven thousand troops from an active role in the battle (Sears, *Gates* 120; Wert 117).

Hours passed as Longstreet and Huger sorted out the mess. It was about noon before Huger's brigades turned onto Charles City Road, and an hour after that before Hill had Brigadier General Robert Rodes's brigade closed on his right flank. Huger's orders had not told him that the battle would not begin until he had relieved Rodes's brigade from its Charles City Road position so that Rodes might join Hill for the attack. Tired of waiting, Hill had sent word to Rodes at 11:00 to come without waiting for Huger, but the waist-deep White Oak Swamp had slowed Rodes's march. Rodes did not reach Hill until 1:00 (Sears, *Gates* 124). Thus some five hours after Johnston had originally envisioned, the Confederate army was in position to begin its assault. This began with Hill advancing through the woods west of Seven Pines until he encountered Federal skirmishers. Behind them were the divisions of Silas Casey and Darius Couch, partially protected by rifle pits and abatis but unprepared for a full-scale attack. In spite of the confused Confederate preliminaries, their advance had not been detected, and they caught the Federals by surprise.

Jeffry Wert concludes that throughout the ensuing battle, "Longstreet did little. He seemingly abdicated his duties as wing commander." After the war, Longstreet admitted in a letter that he "surely had little to do with the battle after the plan of advance and attack was arranged" (Wert 118). Johnston also played an insignificant role in the battle. A freakish combination of atmospheric conditions precluded Johnston from hearing the sounds of the firing from his headquarters, and he would not learn that the battle had been joined for three hours.

After three hours of fighting, Hill requested reinforcements. The planned double envelopment had called for Hill to attack in the center as well as for flanking attacks to occur on both the left and right. The confusion of the morning's events had precluded these planned flank attacks, and now Hill had to create his own. Thus, Longstreet ordered Anderson's brigade forward to assist Hill. Spearheading this attack was

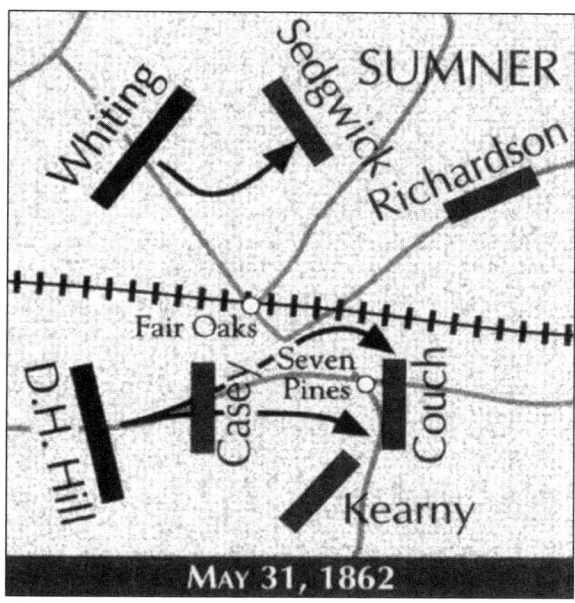

Colonel Micah Jenkins, leading both the Sixth South Carolina and the Palmetto Sharpshooters. Jenkins emerged from the woods on Nine Mile Road south of Fair Oaks and proceeded to rout the Federals, capturing two hundred prisoners and three cannon. Federal resistance around Seven Pines evaporated.

Jenkins's conduct was extraordinary. Captain Thomas Goree of Longstreet's staff wrote that Jenkins "fought 5 separate & distinct lines of the enemy whipping each one." What made Jenkins so successful was that each Federal line that he encountered was of a size he could attack. He had only nineteen hundred men with him, but he was able to locally achieve mass at the decisive place and time in a way that eluded Johnston on the larger battlefield. Jenkins pressed the offensive, and the Federals fell back in disarray.

The results, however, could have been even more dramatic had not Longstreet missed what Confederate artillery officer Porter Alexander called "an opportunity for one of the most brilliant strokes in the war." This would have been to add Pickett's brigade, and possibly Kemper's, to Anderson's reinforcements. Such a force could have rolled up the Federal right flank.

One reason Longstreet missed this opportunity is that he was unaware of it. He did not go forward to assess the situation and was completely unaware of the possibility. Of the 29,500 men in the three divisions under his command, Longstreet managed to get only 12,500 into the battle (Sears, *Gates* 129–40).

The battle of Seven Pines continued for a second day, and in the end the Confederates lost 6,134 killed, wounded, or missing and the Federals 5,031. Johnston accomplished none of his objectives, and after midnight on June 2, the Confederates retreated to the west. One Confederate staff officer described Seven Pines as "a waste of life and a great disappointment" (Wert 123). This failure can be ascribed to three factors: the confusion resulting from Johnston's poor orders, the failure to bring to bear massed forces, and the lack of effective command and control after the battle joined.

Porter Alexander describes Seven Pines as "a monument of caution against verbal understandings" (Wert 123). This statement is extremely relevant today, as commanders rely more and more on technological tools such as video teleconferences in lieu of written orders. Whatever confusion or misinterpretation Longstreet suffered from could have been avoided by a concept sketch, movement overlay, or synchronization matrix. Barring these more advanced tools, a simple back brief in which the commander affirms his subordinate's understanding of his intent by having the subordinate repeat his instructions and explain his plan for executing them would have identified any misunderstanding (FM 101-5 G-1).

Johnston and Longstreet's second failure was that they violated the mass principle of war. The purpose of mass is "to concentrate the effects of combat power at the place and time to achieve decisive results" (Joint Pub 3-0 A-1). Mass requires synchronized action. Johnston had a plan to achieve mass against the two Federal corps isolated south of the Chickahominy, but he was unable to synchronize the movement of his forces required to execute his plan. Longstreet's subsequent orders to Huger further diluted the force. In the larger picture, of the twenty-two infantry brigades Johnston had planned to have assault the Federals, just nine brigades and part of a tenth did all the fighting (Sears, *Gates* 138). Only Jenkins was able to achieve mass by successively attacking piecemealed Federal forces, which, if joined, would have been more than his match.

Underlying both of these first two failures was the ineffective command presence of both Johnston and Longstreet. Johnston's failure resulted largely from selecting a poor location for his headquarters. From that location, he lacked the situational awareness necessary to influence the battle, and actual command therefore fell to Longstreet. In Jeffry Wert's assessment, Johnston "virtually relinquished control of the operation to Longstreet despite knowing as early as 9:00 A.M. that something was wrong. Duty impelled him to ride personally the few miles for an explanation and a possible alteration to the plan to salvage the offensive. Instead, Johnston remained at headquarters, a general wrapped within a cocoon of ignorance and bad luck" (120). Johnston had also chosen to let Longstreet run things at Williamsburg, but there everything was unfolding nicely, and additional direction was unnecessary. At Seven Pines, there was obviously a problem, and Johnston did nothing to correct it.

Longstreet also failed to effectively command at Seven Pines. Like Johnston, Longstreet was not always in the proper position. His limited view of the battlefield precluded him from seeing the opportunity to roll up the Federal right flank by providing a larger counterattack force than just Jenkins's. At Williamsburg, Longstreet, "hearing the rising tide of battle, rode to the front and viewed the scene," but at Seven Pines, like Johnston, Longstreet adopted a passive approach to command (Eckenrode and Conrad 35). He showed none of the personal involvement he had displayed so well at Williamsburg. At Seven Pines, the confusion of the morning's activities, the need for synchronized flank attacks, and the decisions required to get idle Confederate forces into the fight necessitated a hands-on approach to command that Longstreet chose not to provide.

Part of the failure of the Confederate command at Seven Pines can be ascribed to both Johnston's and Longstreet's difficulties with offensive operations. Both had a predilection for the defensive and appeared hesitant to seize the initiative in an offensive role. These traits would reappear later in the war: for Longstreet at Gettysburg and Johnston outside Atlanta.

Seven Pines was Johnston's first offensive of the war. At Williamsburg, he had deferred to Longstreet, likely in recognition of Longstreet's superior tactical skill (Eckenrode and Conrad 36). When Johnston tried to do the same at Seven Pines, he uncovered one of Longstreet's limitations.

H. J. Eckenrode and Bryan Conrad conclude that Seven Pines demonstrated that Longstreet "was as unskillful at offensive tactics as he was
skillful at defensive. Indeed, he had illustrated the essentially defensive
character of his military talent" (49).

Furthermore, Alexander very insightfully identifies the role of the
staff in this process. He writes that Seven Pines "affords a most striking
illustration of how people may misunderstand each other in important
affairs; & of the supreme importance, in such matters, not only of having everything thoroughly understood, but of the commanding general
supervising by his staff the actual execution of all orders in order to
guard against accidents & misunderstandings" (Wert 123).

Military professionals often speak of "command and control" as if
the two functions were one. In reality, command "is the art of motivat-
ing and directing people into action to accomplish missions." Control,
however, is "to regulate forces and functions to execute the commander's intent. . . . Ultimately, it provides commanders a means to measure,
report, and correct performance" (Joint Pub 3-0 II-15). Thus, commanders focus on command, and staffs focus on control. Johnston's staff did
not help him control the operation. There was no system in place to
monitor the morning's movements and make corrections before the situation had gone hopelessly awry. Freeman concludes that "staff work at
General Headquarters on May 30–31 was about as bad as it could have
been" (*Lee's Lieutenants* 1:260).

Having said all this, the most significant event of the day occurred
not on the confused and indecisive battlefield but back at Johnston's
headquarters. After spending most of the battle at a headquarters too
far away to effectively understand the battle, Johnston now decided to
move too close. From a knoll two hundred yards north of Fair Oaks
Station, Johnston was well within the range of Federal fire. After chastis-
ing his staff for ducking as bullets whizzed past, Johnston was struck,
first by a bullet in his right shoulder and then in the chest by artillery
fragments. Gustavus Smith briefly replaced Johnston but "was pros-
trated by the responsibility" (Eckenrode and Conrad 49). At about noon
on June 1, Robert E. Lee arrived on the battlefield with orders from
President Davis to take command.

The challenge facing Lee was immense. Freeman believed that the
"army at that time had been most wretchedly mishandled. A worse executed battle than that of the Seven Pines seldom has been fought in

America. Everything went awry that could go wrong." Nonetheless, the army had considerable devotion to Johnston, and the new commander was met with "a certain patronizing air" (Smith 166). Longstreet writes that "the assignment of General Lee to command the Army of Northern Virginia was far from reconciling the troops to the loss of our beloved chief, Joseph E. Johnston, with whom the army had been closely connected since its earliest life. All hearts had learned to lean upon him with confidence, and to love him dearly. General Lee's experience in active field work was limited to his west Virginia campaign against General Rosecrans, which was not successful. . . . There were, therefore, some misgivings as to the power and skill for field service of the new commander" (Stiles 30–31). The fact that from these inauspicious beginnings the army, by the end of the Seven Days battles, was willing to follow Lee blindfolded is a tremendous testimony to Lee's leadership and one of his greatest accomplishments (Smith 166–67).

THE SEVEN DAYS

After replacing Johnston as commander of the Army of Northern Virginia, Lee quickly took advantage of the opportunity created by Jackson's success in the Valley. On June 8, he wrote to Jackson, "Should there be nothing requiring your attention in the valley so as to prevent your leaving it for a few days, and you can make arrangements to deceive the enemy and impress him with the idea of your presence, please let me know, that you may unite at the decisive movement with the army near Richmond" (War Department, ser. 1, vol. 12, pt. 3, p. 908; Deaderick 34). Three days later, Lee further explained that Jackson would "sweep down between the Chickahominy and Pamunkey, cutting up the enemy's communications, etc., while this army attacks General McClellan in front. He will thus, I think, be forced to come out of his intrenchments where he is strongly posted on the Chickahominy and preparing to move by gradual approaches on Richmond" (War Department, ser. 1, vol. 12, pt. 3, p. 910; Deaderick 34).

A major reason for Lee's desire to unite with Jackson clearly was his paucity of numbers. On the eve of the Seven Days battles, McClellan's force on the Chickahominy numbered 104,300. Lee, even after the arrival of Jackson's command, had just 85,000 in his Army of Northern

Virginia (Sears, *McClellan* 207). Superior intelligence, however, would help Lee make up for what he lacked in numerical strength.

On June 10, Lee had called Jeb Stuart to headquarters and instructed him to conduct a reconnaissance beyond and behind the Federals' right flank, manned by Porter's Fifth Corps. Stuart proposed enlarging the mission beyond reconnoitering the right flank and returning along the same route; instead, he proposed riding entirely around McClellan's army and returning to Richmond by way of the banks of the James River. On June 12, Stuart embarked on this mission, returning to a hero's welcome on June 15.

Stuart brought Lee the information he needed. The Federal right flank was anchored on the north bank of the Chickahominy near Mechanicsville. It did not extend far enough north to block the roads Lee wanted to use to bring Jackson's troops to the battle. Moreover, the Federals' primary supply line, the Richmond and York River Railroad, was vulnerable. By turning McClellan's flank, Lee could threaten the Federal general's communications. On June 16, the day Stuart issued his report, Lee told Jackson, "The sooner you unite with this army the better" (Sears, *Gates* 174).

MECHANICSVILLE

On June 18, Jackson began marching his men toward the Virginia Central Railroad while maintaining the utmost secrecy as to their final destination. It was a 130-mile ride from the Shenandoah to Richmond along the Virginia Central's most direct route. The journey was accomplished by a "riding and tiring" system. A train full of infantry was hauled to a certain location and dropped off. The empty cars were then backed up to the rear of the column, where, as trainman Carter Anderson recalled, "we reloaded the tired ones" and brought them to the head of the column. Jackson arrived at Lee's headquarters at 3:00 in the afternoon of June 23, after completing a fourteen-hour ride on horseback. The rigors of the Valley Campaign combined with the demanding preparations for this new offensive would leave Jackson exhausted and would be a critical factor in his performance during the Seven Days.

Jackson was soon joined at the headquarters by D. H. Hill, Longstreet, and A. P. Hill, and Lee briefed the group on his plan. Whereas Johnston had attacked McClellan's left wing, on the Richmond side of the Chickahominy, Lee's objective would be Porter's Fifth Corps on the Federal right flank. Jackson was to march down from a point due north of Mechanicsville and attack Porter from the flank and rear. A. P. Hill would cross the Chickahominy upstream from the enemy and advance along its north bank, clearing the Mechanicsville Bridge for D. H. Hill and Longstreet to use to join the movement. Magruder and Huger, with a total of 28,900 men between them, would remain on the defensive, guarding Richmond. Theophilus Holmes and his 7,300 men would be in general reserve. After some discussion focusing on Jackson's requirement to make the farthest move, the time of the attack was set for 3:00 in the morning on June 26.

While Lee and his staff were planning, McClellan was active as well. His mind-set was shifting from seeking a single decisive battle to a "series of partial attacks, rather than a general battle." The first of these he planned for June 25 and involved seizing Oak Grove, a woodland in the area of Williamsburg Road and the Richmond and York River Railroad. Control of Oak Grove would facilitate his larger objective of seizing Old Tavern, the high ground on Nine Mile Road a mile and a half in front of the Federal lines.

McClellan received word of Jackson's advance from a Confederate deserter, and McClellan took the precaution of sending out small parties to reconnoiter to the north and to obstruct roads that Jackson might use to reach Porter's flank and rear. Nonetheless, McClellan was not worried enough about Jackson's movements to change his planned attack on Oak Grove. This would become the opening round of the Seven Days.

The battle of Oak Grove began at 8:30 in the morning on June 25, when Joe Hooker sent two brigades due west along Williamsburg Road toward Richmond. Between their jumping-off position and the Confederate lines, a swampy little creek formed the headwaters of the White Oak Swamp and ran through a strip of dense forest some twelve hundred yards wide. It would be difficult terrain through which to maneuver.

The battle consisted of charges and countercharges continuing into the dusk, and darkness ended the inconclusive fighting. In what amounted to a mere struggle over advanced picket lines (McClellan had gained perhaps six hundred yards), the Federals suffered 626 total casualties and the Confederates 441. A combined sum of more than 1,000 casualties for such a small action was a harbinger for just how intense the fighting of the Seven Days would be.

Throughout the fighting at Oak Grove, McClellan continued to receive reports of massing Confederate forces, many of which were Magruder's continued demonstrations. On June 25, McClellan's intelligence chief, Allan Pinkerton, estimated Confederate strength at 180,000. In reality, even including Jackson's forces, which were still en route, it was less than half that figure. These reports sent McClellan into a panic and not only destroyed any offensive intentions he had previously considered but threw him into a spasm of fatalism. James Murfin writes, "Again [Magruder's] deception worked. He was confirming McClellan's intelligence reports" (54). McClellan's thoughts drastically swung from seeking victory to saving his army from defeat, and he began preparing a line of retreat.

McClellan's penchant for always assuming the worst gave him no appreciation for the difficulties facing Jackson and his men. After leaving his meeting with Lee, Jackson rejoined his army at Beaver Dam Creek, just ten miles in advance of where he had left it. He now had two days to assemble his forces and launch the attack planned for 3:00 in the morning on June 26. His attack was to begin at Slash Church, due north of Mechanicsville. To get there he would have to cover twenty-four miles by road in two days—a seemingly manageable task for Jackson's famous foot cavalry. Such, however, was not to be the case.

Jackson had developed an excellent routine for his marches during the Valley Campaign. His men would march for exactly fifty minutes, halt for a ten-minute rest, and then resume the march. In the middle of the day, at 12:00 or 1:00, they would have an hour for lunch. The role of the chain of command was clearly articulated: "Brigade commanders will see that the foregoing rules are strictly adhered to, and for this purpose will, from time to time, allow his [sic] command to move by him, so as to verify its condition. He will also designate one of his staff officers to do the same at such times as he may deem necessary" (Freeman, *Lee's Lieutenants* 1:370).

However, Jackson had never directed an army as large as he did now, and he could not be everywhere at once. Units beyond Jackson's personal reach tended to straggle, and his staff did little to help control and facilitate the movement. The situation was worsened by continued rain that made the roads muddy and the creeks high. By the end of June 25,

Jackson was still five miles short of Slash Church. Nonetheless, he sent a courier to inform Lee that Jackson planned to begin his march at 2:30 the next morning and would be in position at the agreed-upon time to begin the attack.

Jackson, however, would not be moving at 2:30. Instead, it would be nearly 5:00 before his columns started marching. Jackson was a full five miles from Slash Church and two and a half hours late in starting. To turn Porter out of his line behind Beaver Dam Creek, just east of Mechanicsville, Jackson had an eighteen-mile march ahead of him (Sears, *Gates* 174–76, 181–95).

The terrain around Beaver Dam Creek favored the defense. The creek itself flows south through the battle area and empties into the Chickahominy. On either side of the creek lie eighty to one hundred yards of marshland and thick undergrowth. Such an obstacle was passable for infantry, but cavalry and artillery would require a bridge. Two

Bridge over Chickahominy River on Mechanicsville Road

were available—the Meadow Bridge and the Mechanicsville Bridge. The Meadow Bridge was about a mile and a half upstream, leaving the Mechanicsville Bridge as the better option for the Confederates.

Both sides of Beaver Dam Creek hosted ridge lines. The eastern one, although slightly lower than the western one, afforded observation and fields of fire beyond Mechanicsville (Glass 11). The Confederates had recognized this key terrain, planning to occupy it when Johnston was intending to defend north of the Chickahominy earlier in May. When Johnston instead fell back behind the river, Confederate artilleryman Porter Alexander lamented that "the enemy took the beautiful Beaver Dam position for his own right" (Sears, *Gates* 194).

In charge of these Federal positions was Brigadier General George McCall, accompanied by a ninety-five-hundred-man division of Pennsylvania Reserves. McCall positioned outposts at Meadow Bridge and Mechanicsville but instructed them to fall back to the main defenses if attacked. Porter wisely intended to fight only from behind Beaver Dam Creek. He explains that the small outposts at the bridge sites "were conspicuously displayed for the purpose of creating an impression of numbers and of an intention to maintain an obstinate resistance. We aimed to invite a heavy attack, and then, by rapid withdrawal, to incite such confidence in the enemy as to induce incautious pursuit" (Wheeler 297).

McCall put John Reynolds's and Truman Seymour's brigades in the main line and held George Meade's brigade in close support. He was dug in securely in a mile and a half of prepared positions. Key to the defense were thirty-two guns in six batteries trained on the clear fields of fire around the creek. To enhance the natural obstacle, McCall's men had built an abatis on the west bank. Alexander described the position as "absolutely impregnable to a frontal attack" (Sears, *Gates* 202). Unfortunately for the Confederates, a frontal attack is exactly what would transpire.

Such, of course, was not Lee's intention. He had designed a turning movement to force the Federals out of their positions, telling President Davis that the "experiment" of frontal attacks Lee had witnessed at Seven Pines would not be repeated. Lee's plan was sound enough, but Clifford Dowdey points out that its critical vulnerability was that "it depended on the human element" (*Land* 190).

The chief failings in this department were those of Jackson, whose delays foiled Lee's plan. It was 9:00 A.M. before Jackson reached Slash

Church on the Ashcake Road. From there he advanced by two roads leading south and converging at Hundley's Corner. Here the plan was for A. P. Hill's division to join him on his right and for D. H. Hill's division to come up in support, but by now it was 5:00 P.M. Moreover, Jackson had provided Lee with no information regarding the situation. Jackson had notified Lawrence Branch, one of A. P. Hill's brigadiers, of the delay at 10:00 A.M., but that was his only communication with anyone in Lee's army throughout the day. In another human failure, Branch, a North Carolina legislator and undisputed patriot but who lacked any military experience whatsoever, failed to pass that message to Lee (Dowdey, *Land* 190–92).

While Branch can certainly be faulted for failing to inform Lee of Jackson's delay, a more dynamic and redundant form of communication and coordination between Jackson and Hill was clearly called for. Stuart's entire cavalry had been sent to screen against an attack from the east, but hindsight suggests that a small detachment of mounted couriers might have been used to great affect to facilitate communication between the separate wings (B. Alexander 99).

In the absence of such communication, A. P. Hill had grown restless and taken matters into his own hands. At 3:00 P.M., having heard nothing of Jackson, Hill sent his division across Meadow Bridge, scattering McCall's pickets, and prepared to advance on Mechanicsville. The Federals at Mechanicsville followed Porter's plan and withdrew to their lines behind Beaver Dam Creek.

Lee caught up with Hill in Mechanicsville and learned that Hill too knew nothing of Jackson and had acted alone. This was not what Lee had envisioned. Jackson's planned march against Porter's right rear had been intended to turn Porter out of his strong position and make him vulnerable to an attack by combined Confederate forces. Now, however, with Jackson's whereabouts unknown and Hill already in contact, Lee, as he explained later, felt "obliged to do something." To halt now would pass the initiative to the enemy. McClellan could then strike the weakened Richmond or attack Jackson's isolated corps. Lee decided to press the attack (Sears, *Gates* 194–203).

In so doing, Hill launched four separate assaults against the formidable Beaver Dam Creek positions. All were frontal attacks across the exposed and marshy terrain with little or no support "against a position the strength of which was so well known that a turning movement upstream had been regarded as the *sine qua non*" (Freeman, *Lee's*

Lieutenants 1:516). All were subjected to a murderous fire from the Federal artillery on the ridge. In the words of Freeman, "All the Federals had been called upon to do was to wait for the Confederates to come into range and then mow them down" (*Lee's Lieutenants* 1:515). The attacks ended with nightfall around 9:00. The Confederates lost more than fifteen hundred dead and wounded. The Federals lost fewer than four hundred.

Very late in the day, the Federals had discovered Jackson and believed that he threatened both the supply line and the Beaver Dam Creek position. Rather than taking advantage of the Confederates' failed attack, separated forces, and weakened capital, McClellan decided to retreat (Glass 13–14). In the words of Stephen Sears, McClellan "surrendered to defeat almost as soon as the fighting began" (*McClellan* 286). Lee had lost the battle but was on his way to winning the campaign.

At Mechanicsville, Lee had faced the same problem that had thwarted Johnston at Seven Pines—the need to carefully synchronize the movement of his forces to gain a local advantage. Instead of the 55,800 men Lee had planned to throw against the Federal flank, barely 11,100 had gotten into action, and these were employed frontally.

In another unfortunate similarity with Johnston at Seven Pines, Lee may have suffered from a failure of a subordinate—in this case, Jackson—to understand his superior's intent. Freeman allows that Lee's orders for the attack were not entirely unambiguous, and A. P. Hill's biographer, Martin Schenck, explains that "Jackson *could* have interpreted them as being merely marching orders, directing him to Cold Harbor. [It] is possible that he thought that Lee, being on the ground, would give the word when the attack was to be launched, after the situation had developed to the point where such an attack would enjoy the greatest success" (61). If Schenck is correct, the situation might have been avoided by the same back brief procedure that might have helped Johnston ensure that Longstreet understood his intent at Seven Pines. Schenck also notes that Jackson asked no questions at the June 23 conference with Lee, thus implying that Jackson thought he understood Lee's intent (62). It is the commander's responsibility to ensure that the subordinate does in fact understand, and the back brief is the best vehicle for doing so.

Even if the Confederates had overcome these difficulties, the terrain around Mechanicsville may have made the Federal defense still too tough to be beaten. Using the OCOKA mnemonic, the strength of Porter's position is readily appreciated. The observation and fields of fire afforded by the ridge east of Beaver Dam Creek were outstanding. The Federals could easily observe the approach of the Confederates and held fire until Hill's men were trapped in the swampy quagmire around the creek. Additionally, Porter reports being able to see "vast clouds of dust" to the north and west, which indicated to him that Jackson was still far in the distance and posed no immediate threat (Wheeler 297).

While the Federals enjoyed cover and concealment up on the ridge, there was no such protection for the Confederates in the lowlands. Porter Alexander observed that "there was no cover in front within musket range, say 400 yards, & the enemy's line could not be reached by an assaulting force, & his men were quite well sheltered from fire" (E. Alexander 99).

The creek itself was a significant obstacle. Alexander continues by saying that the "valley of the creek was rendered impassable by the fallen trees & brush, & by the creek on one side of it, & the mill race on the slope of the eastern bank, just in front of the enemy's line" (E. Alexander 99). Such conditions created an excellent engagement area. Porter explains that "the banks of the valley were steep, and forces advancing on the adjacent plains presented their flanks as well as their front to the fire of both infantry and artillery, safely posted behind entrenchments. The stream was over waist-deep and bordered by swamps" (Wheeler 297).

These conditions would canalize the Confederates along the Meadow or Mechanicsville Bridges, which, along with the ridge itself, represented key terrain. Concerning the ridge, it is interesting to remember that although the western ridge was higher, the eastern one dominated the battlefield area and therefore was key terrain.

That the avenue of approach would be restricted to the bridge was made even more dangerous for the Confederates by the fact that part of their route would run parallel to the Federal line, thus leaving them extremely vulnerable to Porter's artillery (Hassler, *McClellan* 143). What made these characteristics especially dangerous is that they were not obvious to the attacker. Alexander concludes that "the full strength of the position, particularly the inaccessible feature of it, was not apparent to the eye until one had entirely crossed the plain swept by their fire & gotten actually up to the valley of the creek." The Federals allowed Hill to advance confidently into this trap and then cut him down (E. Alexander

99–100). At Beaver Dam Creek, the terrain unquestionably favored the defense, and the Federals took full advantage of it.

Mechanicsville was Lee's first battle in command of the Army of Northern Virginia. Through brilliantly conceived, it proved to be too difficult to execute. The visible reason that Lee's plan failed was Jackson's late arrival. The underlying cause, however, was that the plan violated the simplicity principal of war. Of this Freeman writes, "It was the high command that had failed. None could deny that. Except for [two instances], no detail of the plan of action had been executed on time and in accordance with the plan" (*Lee's Lieutenants* 1:515–16). This problem of execution and synchronization would plague Lee throughout the Seven Days.

GAINES' MILL

When McClellan decided to retreat, crossings over the swampy terrain became critical to him. He ordered Porter to pull back to a defendable position covering the Chickahominy bridges. John Barnard, McClellan's chief engineer, laid out the defense to cover the bridges. The general area was a largely open, oval-shaped plateau, varying in height from forty to eighty feet. The highest elevation was known locally as Turkey Hill, although the battle would be named for Gaines' Mill, a full mile away.

Beginning at the northeastern corner of the plateau and curving around its northern and western sides before emptying into the Chickahominy was Boatswain's Swamp. Its banks and bottomlands were heavily overgrown, and toward its mouth it was steep sided and especially marshy. Boatswain's Swamp was not the obstacle Beaver Dam Creek was but would prove to be rough going.

To the north and west of the plateau, the ground was largely open and sloped down toward the swamp. On the Federals' side, the land rose more steeply. Porter placed his corps in a mile-and-three-quarter crescent facing north and west. George Morrell's division was on the left, or western, flank, and George Syke's was on the right, or northern, flank. The men were arranged in two lines, one near Boatswain's Swamp and the other halfway up the hillside. McCall's division comprised a third line, the reserve, at the crest of the plateau.

As at Beaver Dam Creek, Porter had found excellent defensive terrain and used it to is advantage. He wrote that the "west bank gave excellent protection to the first line of infantry, posted under it, to receive the enemy descending the cleared field sloping to it. The swampy grounds along the sources of the creek were open to our view in front of hundreds of yards, & were swept by the fire of infantry & artillery. The roads from Gaines' Mill & Old Cold Harbor, along which the enemy were compelled to advance, were swept by artillery posted on commanding ground" (E. Alexander 101). Once again, Porter had all the factors of OCOKA working in his favor.

In all, Porter had 27,160 men in position for the battle of Gaines' Mill, but, as at Beaver Dam Creek, artillery would play the key role. Seventeen Federal batteries—ninety-six guns in all—were positioned in line or in reserve across the plateau. Additionally, three batteries from Franklin's Sixth Corps south of the river could range any assault against Porter's left flank. Freeman concludes, "In a word, the Federals had a perfectly protected position, of which they had made the most" (*Lee's Lieutenants* 1:520).

Lee knew that McClellan was defensive minded but did not suspect the magnitude of McClellan's response to Mechanicsville. Lee believed that rather than completely abandoning the campaign, McClellan would merely reposition to defend his White House supply line. Thus, Lee intended to keep up the pressure, and he ordered A. P. Hill to move across Beaver Dam Creek after learning that Porter had abandoned his position. Lee believed that the Federals would withdraw to Powhite Creek, the next good defensive position.

When Lee finally made contact with Jackson at 10:30 A.M. on June 27, Lee formed a plan for an envelopment. Jackson was to march northeast on the Old Cold Harbor Road across the headwaters of Powhite Creek to Old Cold Harbor, where he would be joined by D. H. Hill. Hill was already conducting his own wider turning movement over the Old Church Road further north. When joined, Jackson and Hill would have fourteen of the Army of Northern Virginia's twenty-six brigades. While A. P. Hill and Longstreet kept the Federals busy from the front, Jackson and D. H. Hill would threaten the enemy communications with the Richmond and York River Railroad. Huger and Magruder would be left to defend Richmond, a task that would give Magruder another opportunity to practice his skills as the Great Demonstrator.

As it would turn out, Lee would face some of the same problems at Gaines' Mill that he had faced at Mechanicsville. At Mechanicsville, A. P. Hill had initiated the first contact on his own. Similarly, at Gaines' Mill, D. H. Hill would reach Old Cold Harbor well before Jackson and find himself prematurely joined in battle between noon and 1:00 P.M. Hill, however, had run into a larger force than he had expected. Furthermore, he appeared to be at the enemy's front rather than its flank. He decided to wait for Jackson. As at Mechanicsville, however, Jackson would be painfully slow in arriving.

While D. H. Hill was waiting, A. P. Hill had advanced down Telegraph Road toward Gaines' Mill and New Cold Harbor beyond. At New Cold Harbor, A. P. Hill's lead brigade under the command of Maxcy Gregg encountered a severe Federal artillery barrage. Lee and Hill quickly moved to Gregg's position and learned that the Federals had not made their stand at Powhite Creek but had occupied a considerably stronger position further east at Boatswain's Swamp, a feature that did not appear on Lee's map.

To meet this new situation, Lee ordered the rest of A. P. Hill's division and that of Longstreet to move up and form a line. Porter could observe these developments from his headquarters on the hilltop and requested reinforcements. Henry Slocum's division from Sixth Corps was sent forward to cross the Chickahominy at Alexander's Bridge.

At 2:30, A. P. Hill gave the order to advance. Most of his men would have at least a quarter of a mile of open ground to cross before they reached Boatswain's Swamp. They would be facing three batteries of Federal artillery posted on the lower slopes of the plateau and several higher up on the crest. For nearly two hours, Hill's men struggled against these odds and got nowhere. The entire day, Hill would lose 2,154 men, and most of these losses would occur right here. For Hill, it was Beaver Dam Creek all over again. He was taking a furious pounding in a frontal assault across open ground into Federal artillery. Moreover, Jackson was nowhere to be found.

In fact, Jackson had obtained a guide at Walnut Grove Church and apparently told him only that the general wanted to go to Old Cold Harbor. By what Freeman attributes to "soldierly reticence," Jackson did not impress on the guide the need to arrive at Old Cold Harbor by a route that would approach the Federals from the flank (Freeman, *Lee's Lieutenants* 1:524). Not until three miles into the march, just short of

Powhite Creek, did the guide discover Jackson's intent. At that point, there was no choice but to backtrack to Old Cold Harbor Road. By the time Jackson's lead division under Dick Ewell reached Old Cold Harbor, Lee's aide, Walter Taylor, was there to meet Jackson. Lee was concerned by A. P. Hill's vulnerability to a counterattack and had sent Taylor to direct Jackson's men to the battlefield. Ewell hurried his brigades into action and ended up committing them piecemeal. The Federals defeated each in turn.

By now, Slocum's reinforcements were reaching Porter, and Porter would use them to plug gaps in his lines. He was under increasing pressure as Longstreet was now launching diversionary attacks on the right to try to afford some relief for Hill. Porter reported to McClellan, "I am pressed hard, very hard"; unless reinforced, "I am afraid I shall be driven from my position" (Sears, *Gates* 233).

Just as at Yorktown, McClellan had no branch planned for meeting this situation. Instead of acting, he noncommittally asked his subordinates if they had any troops to spare. Sears concludes, "Awash in his own anxieties, General McClellan had given no thought to anticipating Porter's need" (*Gates* 233). Sears correctly assesses that McClellan characteristically spent time wringing his hands when he should have been planning branches. In the end, McClellan responded by sending Porter two brigades from Sumner's Second Corps. This was a mere one-tenth of the forces available to McClellan on the south side of the Chickahominy, where he was being paralyzed by nothing more than Magruder's theatrics.

While McClellan was so parsimoniously reinforcing Porter, Jackson's belated arrival was providing Lee with additional men. Daylight was beginning to fade, and there was time for just one more large-scale attack before darkness. Both Lee and Jackson had gravitated to the center of the battlefield and ultimately met on Telegraph Road. Lee said, "Ah, General, I am very glad to see you. I had hoped to be with you before." Jackson only nodded at the gentle rebuke.

The result of the meeting was that Jackson would add Chase Whiting's and Charles Winder's divisions to the fight, attacking on Longstreet's left. After combat-ineffective units were removed and two of Winder's brigades and one of Longstreet's were allocated to the reserve, Lee had 32,100 men in sixteen brigades to throw against Porter's 34,000 remaining effectives.

The climax of the battle of Gaines' Mill was a confused melee. D. H. Hill advanced with his five brigades on the left. To his right was

Ewell, then Whiting, then Longstreet. In the words of Stephen Sears, it "was a battle fought entirely without subtlety; the tightly contained Federal position appeared to offer no opportunity for maneuver. The result was a straight-ahead slugging match in which the defenders held all the advantages of position" (Sears, *Gates* 240).

However, Confederate persistence ultimately carried the day. John Hood's brigade of Whiting's division is traditionally credited with being the first to break the Federal line. Indeed, Lee had earlier sought out Hood, explained the situation to him, and said, "This must be done. Can you break this line?" Hood promised, "I will try" (Sears, *Gates* 240). On Whiting's instructions, Hood and Evander Law advanced their brigades without pausing to fire, covering the open ground down to Boatswain's Swamp as quickly as possible. Hood had split his men, making the charge on both flanks of Law's brigade.

Hood and Law suffered staggering casualties. Between them, they lost 1,018 men at Gaines' Mill, at least two-thirds of them from this charge. Nonetheless, the Federals, able to fire at best three shots a minute, were unable to keep up with the swift pace of the attack. The Federal line broke.

It was soon the same everywhere. Almost simultaneously, Porter's defense cracked. There was no choice now but to retreat toward the Chickahominy crossings. Darkness covered the move.

Gaines' Mill would prove to be the largest and most costly battle not just of the Seven Days but of the entire Peninsula Campaign. A total of 96,100 men had been on the field. In less than nine hours of fighting, Porter had suffered 6,837 total casualties and Lee 7,993. The Federal failure is best explained because McClellan did not fight the battle to win. He had left 64,000 men idle on the south side of the Chickahominy while Porter fought alone. Although victorious, the Confederates suffered high casualties, which also can be explained by an inability to get forces into the fight.

The loss caused McClellan to announce to his lieutenants what he had privately already decided: he would abandon the campaign and shift his base from White House to Harrison's Landing, where he would be under the protection of Flag Officer Goldsborough's gunboats (Sears, *Gates* 210–50). Furthermore, McClellan sent a long dispatch to Secretary of War Stanton that closed, "If I save this army now, I tell you plainly that I owe no thanks to you or to any other person in Washington. You have done your best to sacrifice this army." Even McClellan's sympathetic biographer, Warren Hassler, confesses that the dispatch was "disrespectful and insubordinate" (Hassler, *McClellan* 153). As such, it clearly indicates the magnitude of McClellan's contempt for the Lincoln administration, but it also shows the general's defeatist and fatalistic state of mind at this point in the campaign.

SAVAGE'S STATION

With McClellan's decision to withdraw, Federal forces began streaming across the Chickahominy. Federal batteries controlled the crossings from the south, preventing Lee from pursuing. In contrast to the rest of the Seven Days, June 28 would pass quietly.

George McClellan's headquarters at Savage's Station

Lee considered that McClellan might be following any of three courses of action. The first was to remain on the Chickahominy and fight to defend his railroad supply line to the White House. The second would be to withdraw down the Peninsula, recross the Chickahominy further downstream, and regain his link with the York River. The third possibility was to abandon the Chickahominy line altogether and retreat south to the protection of the Federal navy on the James. To ascertain which one of these courses McClellan would follow, Lee sent out both Dick Ewell's division and Jeb Stuart's cavalry.

These reconnaissance missions soon reaped benefits. By midafternoon, Stuart had sent word that he and Ewell had found the Richmond and York River Railroad station undefended. As the Confederates pushed on to

Ruins of the White House after it was burned by withdrawing Federal troops

the Chickahominy crossings, Federal guards set the railroad bridge and Williamsburg Road crossing at Bottom's Bridge on fire. McClellan was clearly abandoning his railroad supply line. Then, when Stuart reached White House, he found it evacuated and a great many supplies in flames. Lee now knew for sure that McClellan was in full-scale retreat. Thus, Lee set out to intercept the fleeing Federals. To do so, he would use the network of four roads that fanned out east and south from Richmond.

Lee assigned the immediate pursuit mission to John Magruder, who would move due east along Williamsburg Road and the Richmond and York River Railroad in hopes of catching the Federals' rear guard and forcing it to turn and fight. Benjamin Huger would advance on Magruder's right along Charles City Road to overtake the Federals south of White Oak Swamp at Glendale on Quaker Road. Stonewall Jackson, with D. H. Hill still attached, had the shortest route. He would move due south, cross the Chickahominy by rebuilding the Grapevine Bridge, and close with the Federal rear guard in the vicinity of Savage's Station. Thus,

Grapevine Bridge

Lee had a three-pronged pursuit to catch the Federals with Magruder in the center, Huger on the right, and Jackson on the left. Lee would use the rest of his army to intercept the Federals.

Of these intercepting forces, Longstreet and A. P. Hill had the longest marches. They were assigned Darbytown Road on Huger's right. By moving on this route, they too could converge on the Federals at Glendale. Theophilus Holmes would follow the southernmost of the four roads from Richmond, River Road, which reached the James River at Malvern Hill. Thus, Lee had planned an immediate pursuit designed to catch and slow the Federals and a broader pursuit designed to block and exploit.

As had all his plans for the Seven Days thus far, Lee's current plan had many moving parts. Synchronization and timing would be critical. As before, Stonewall Jackson was a key player.

In another scene familiar to the recent fighting, McClellan had absented himself from the immediate battlefield and left no one in charge. Sumner's, Franklin's, and Heintzelman's corps comprised the rear guard, with each commander doing what he thought best. As a consequence, Heintzelman and Sumner withdrew independently of each other, leaving Sumner's flank uncovered (Sears, *McClellan* 217). Such was the condition when Magruder caught up with Sumner at 9:00 A.M. on June 29 at Allen's Farm, two miles short of Savage's Station.

Magruder had been up all the previous night. The effects of this exhaustion were so great that Freeman titles this chapter of *Lee's Lieutenants*, "Magruder Stays Up Too Long" (1:538). Moreover, Magruder was suffering from acute indigestion, and his surgeon had treated him with a mixture that contained morphine. Always nervous, these added stimuli put Magruder especially on edge this day. He would not be up to the task before him.

Indeed, Magruder convinced himself that instead of retreating, the Federals were preparing to attack him. He sent word to Lee that his forces had "found the enemy in numbers far exceeding" theirs and requested reinforcements. Magruder was correct that he was outnumbered; to his 14,000 men, Sumner had 26,600. Nonetheless, Lee was incredulous that the fleeing Federal army would now halt in its tracks and go on the offensive. In spite of his chagrin, Lee sent Magruder two of Huger's brigades under the condition that they would be returned at 2:00 P.M. if they were not engaged (Freeman, *Lee's Lieutenants* 1:550–51). Magruder would not use this force.

When 2:00 came, Huger arrived at Magruder's headquarters, noted the quiet, and reclaimed his men. Shortly thereafter, Magruder received word from Jackson that he would be unable to cooperate with Magruder because "he has other important duty to perform." Magruder was left to his own devices, and at 5:00 P.M., he cautiously engaged the Federals at Savage's Station.

As Magruder advanced, his lead brigade of Joseph Kershaw was spotted by the Federals. The forces were soon engaged in a fierce firefight. However, the fighting devolved into a stalemate, with both Magruder and Sumner showing marked caution.

Sumner's Second Corps consisted of twenty-six regiments, but he committed only ten at Savage's Station. Of the six brigades Magruder had available, he used only two and a half. In Magruder's mind, he was accomplishing his mission of forcing the Federal rear guard to turn and fight. He would just maintain contact and his position and wait for Jackson. In light of such caution, Freeman writes, "Forgotten, apparently, were Magruder's orders to pursue vigorously" (*Lee's Lieutenants* 1:551).

To make matters worse, Jackson again would not be forthcoming. The "important duty" to which he had referred earlier had turned out to be the result of a miscommunication. Lee's chief of staff had sent Jackson a message informing him that Stuart's cavalry would guard the Chickahominy crossing points, "advising Gen'l Jackson, who will resist their passage until reinforced." Jackson interpreted this as requiring him to remain in place north of the river, ready to defend the crossings and prevent an enemy escape if Stuart's cavalry reported any such activity. What Lee had intended to be a mere contingency as part of the original plan, Jackson had concluded to be his new primary mission.

Even without this mix-up, Jackson ran into trouble at the Grapevine Bridge, a ramshackle affair erected by Sumner's troops back in May. It had seen little use since Federal engineers under Lieutenant Colonel Barton Alexander had built a far superior bridge, now destroyed, four hundred yards upstream that bore Alexander's name. The Grapevine Bridge was repaired by noon, but it was soon deemed insufficient to rapidly transport Jackson's five divisions. Thus, Jackson put his chief of staff, Major Robert Dabney, in charge of rebuilding Alexander's Bridge. Dabney was a preacher by trade, not an engineer, and the task proved too much for him to handle. After realizing this, Jackson placed an experienced engineer, Captain C. R. Mason, in charge of the job, and Mason had the bridge complete that night. With this delay, Jackson did not arrive at Magruder's headquarters until 3:30 A.M. on June 29.

Of course by that time, it was too late for Jackson to be a factor at Savage's Station. The fighting had ended at about 9:00 the previous night, with both sides holding pretty much what they had held at 5:00. Later in the darkness, the Federals would withdraw across White Oak Swamp.

In the combined fighting at Allen's Farm in the morning and then at Savage's Station, the Federals lost 1,038 men and the Confederates only 473. Nonetheless, it was a disappointing day for General Lee. He had failed to cut off the Federal retreat.

Lee was very critical of Magruder, telling him, "I regret much that you made so little progress today in the pursuit of the enemy. In order to reap the fruits of our victory the pursuit should be most vigorous. . . . We must lose no more time or he will escape us entirely." Nonetheless, in spite of Magruder's caution, it is easy to understand his actions. He was heavily outnumbered, and it must have seemed prudent for him to wait for Jackson. Lee's ire would seem to have been better reserved for Jackson, who had once again failed to arrive at the decisive point and time (Sears, *Gates* 256–76).

GLENDALE

Lee, however, was not willing to abandon his pursuit because of this failure to bring it to its desired conclusion at Savage's Station. On June 30, Lee met with Jackson and placed him in charge of the direct pursuit of McClellan. Magruder would return to Darbytown Road and form a reserve for Longstreet and A. P. Hill.

The key terrain in the operation was Glendale, a small crossroads community two and a quarter miles southwest of White Oak Bridge. Glendale was key because all roads, including Long Bridge Road, Charles City Road, Darbytown Road, and Quaker Road, led to or toward it. The Federal forces, whichever White Oak Swamp crossing they used, had to funnel through Glendale; from there, most would follow Quaker Road to the James. For this reason, McClellan would have to fight to defend Glendale.

Lee's plan was to throw the largest part of his army, six divisions totaling 44,800 men, directly against Glendale to cut McClellan's army in two. Simultaneously, Jackson and his four divisions of 25,300 men would engage the Federal rear guard. The only forces Lee would not be putting into the fight were Holmes's division and Stuart's cavalry.

Lee had developed a plan to mass his forces in the pursuit. The retreat, conversely, had left the Federal army, in spite of its overall superior numbers, widely scattered. The result was that locally at Glendale and White Oak Swamp, Lee could outnumber McClellan 71,000 to 61,500. Such odds moved Porter Alexander to recollect later that there were but a handful of days during the Civil War in which the Confederates "were within reach of military successes so great that we might have hoped to end the war with our independence. . . . This chance of June 30 '62 impresses me as the best of all" (E. Alexander 110).

With the Federal army in such a precarious state, one might expect McClellan to provide personal leadership and example, but such was not the case. Instead, he established a safe, comfortable, and completely irrelevant headquarters aboard the *Galena* on the James. Such a decision is a marked contrast to both General Lee and President Davis's being so far forward on the battlefield that A. P. Hill had to order them to the rear (Schenck 83–84).

Stephen Sears explains McClellan's behavior by concluding that he "had lost the courage to command." The steady, demoralizing effects of the Seven Days had overwhelmed McClellan, and by this point "exercising command in battle was now quite beyond him, and to avoid it he deliberately fled the battlefield" (Sears, *Gates* 281). Moreover, as he had done previously, McClellan failed to put anyone else in charge when he absented himself. Consequently, as Sears puts it, "Each corps commander, and sometimes each division commander, had placed his men as he saw fit, and consequently it was far from being a seamless line of battle. Some units were advanced, some drawn back, and there were numerous gaps. It was a rare Federal general that day who knew the identity of the units on his flanks" (Sears, *Gates* 282).

Against this confusion, Lee had Huger's 12,000-man division pursuing down the Charles City Road. Huger had been thus far only lightly engaged during the Seven Days, and his division was now Lee's largest. Huger also had the day's shortest march to Glendale, just three miles. Longstreet would command both his division and A. P. Hill's and would march seven miles to Glendale via Darbytown and Long Bridge Roads. Longstreet's combined command numbered 19,200 men, and he also had Magruder's 13,600 men in reserve. Jackson's four divisions of 25,300 men oriented on the White Oak Swamp crossings, where the Federal rear guard was thought to be located. Lee positioned himself in the center with Longstreet and Hill and waited for Huger to initiate the battle.

Lee would have a long wait. Huger was a former ordnance officer, and, like so many generals early in the war, he now found himself in a field command beyond the limits of his abilities. Rather than vigorously pursuing, he advanced cautiously, like Magruder somehow expecting the fleeing enemy to turn and attack him at any moment. Thus, it took Huger all morning to advance just over a mile. The Federals took advantage of this gift of time to leave a series of abatis blocking the Confederates' path, and rather than merely lifting the logs out of the way, Huger allowed himself to be convinced to cut a new road through the woods. Freeman describes the result as "a battle of rival axemen" (*Lee's Lieutenants* 1:568).

At this slow pace, Huger did not even get within artillery range of Brigadier General Henry Slocum's division of Franklin's Sixth Corps until late in the afternoon. Then, rather than closing with superior numbers of infantry, Huger cautiously contented himself to engage in an artillery duel. It apparently never occurred to him to leave his guns in the rear and keep up an active skirmish by pressing forward with his infantry through the woods (Freeman, *Lee's Lieutenants* 1:568). Lee would get little help from Huger this day.

With Huger inching along in this manner, Jackson continued to have problems of his own. When he reached White Oak Bridge, he found it burned, and the marshy bottomland at the site precluded any crossing of his wagons and artillery. He also observed several Federal batteries, supported by infantry, across the stream. These Federals, however, did not see Jackson, and Jackson silently deployed his own artillery on line. At about 2:00 P.M., his artillery opened fire.

In the Mexican War, Jackson had seen the offensive breaching power of close-in artillery at Chapultepec. Frank Vandiver notes that in the midst of the Civil War, Jackson began looking for a way to re-create this power by having massed artillery clear the way for infantry (314). Jackson now had his chance to test his hypothesis. His artillery chief, Colonel Stapleton Crutchfield, had deployed seven batteries of twenty-three guns for a diagonal line of fire that caught the Federals by complete surprise. The bombardment indeed shattered the Federals' artillery, but the main body of their infantry remained unshaken. Jackson ordered a battery slipped close to the bridge site, directing it drive off the

Federal sharpshooters while Colonel Tom Munford's Second Virginia Cavalry forced a crossing. It would be the same use of artillery as had carried the day at Chapultepec.

Munford, however, was quickly repulsed in what Vandiver calls "a lesson in futility" (314–15). Artillery no longer held the close-in power over the infantry it had enjoyed in Mexico. The longer ranges the rifle now afforded the infantry had negated the artillery's advantage in breaching infantry lines (Weigley 90). Jackson learned a lesson in the impact of technology that remains applicable. Today's defenders benefit from "the increasing range and precision of direct and indirect fires [that] allow Army forces to weaken attackers and shape the situation before entering close combat" (FM 3-0 8-18). Technological advances in the rifle would make the Civil War's most notable artillery successes occur on the defensive, as at Malvern Hill, whereas in Mexico such successes had been offensive.

In the midst of all this, Munford did discover a cow path crossing site just a quarter mile downstream. There were no Federals at the site, and Munford sent word to Jackson that an infantry crossing there could strike the enemy's flank. Jackson did not reply.

Similarly, Wade Hampton, commanding a brigade in Charles Winder's division, soon discovered a second potential crossing site to the east of White Oak Bridge. Hampton reconnoitered the site and found not only that the Federals were not covering it but also that from it the Federal eastern flank could be turned. He reported the information to Jackson, who instructed Hampton to build a bridge for the infantry.

Hampton quickly had his bridge completed and reported so to Jackson, finding the general seated on a log with his cadet cap pulled low over his eyes and bearing a completely passive expression. Hampton volunteered his brigade for the movement and waited for orders, but, according to Hampton, Jackson only "sat in silence for some time, then rose and walked off in silence." Jackson ultimately would go to sleep under a tree. His opponent, Franklin, later wrote, "We should have been defeated that day had Gen. Jackson done what . . . he should have done" (Schenck 114–15). In yet another instance of inactivity and slowness, Jackson would fail Lee during the Seven Days.

While Huger and Jackson stagnated and Lee held up action waiting for them to proceed as planned, Colonel Thomas Rosser of the Fifth Virginia Cavalry had been scouting the army's right flank. Shortly after

3:00 P.M., he reported to Lee that he had seen Federal columns "moving hurriedly and confusedly" over Malvern Hill. Lee rode off to find Theophilus Holmes and see what he knew of the situation. Lee found Holmes calling up his artillery and preparing to shell the enemy. Lee instructed Holmes to bring up his entire division and then returned to the Glendale front.

Now Lee knew that the head of McClellan's retreating army was at Malvern Hill and the tail was at Glendale and White Oak Swamp. He determined to strike both simultaneously. Thus, he ordered Magruder's divisions, currently in reserve at Glendale, to march to support Holmes's attack on Malvern Hill. This decision seemed sound enough because Lee was counting on Jackson and Huger bearing a significant portion of the fighting at Glendale. In reality, not a single one of Huger or Jackson's infantry would fire a shot. Thus, by dispatching Magruder, Lee had inadvertently severely weakened his force at Glendale.

As Magruder's command maneuvered, Lee waited and waited at Glendale for Huger and Jackson to act. Lee finally could wait no longer and ordered Longstreet to attack. Longstreet organized his attack into an initial assault line of three brigades led by Richard Anderson. Behind them were three brigades in close support. Lawrence Branch's brigade of A. P. Hill's division was supporting on the right flank, and the rest of Hill's division remained in reserve. On the Federal side, George McCall's division stood directly in front of the Confederate advance, with Phil Kearny's division to the north, Joe Hooker's division to the south, and a brigade of John Sedgwick's division in the rear. There was nothing to even approximate unity of command over these Federal forces.

Stephen Sears describes the resulting battle at Glendale as "a battle of reinforcements. . . . It was a matter of numbers" (Sears, *Gates* 300–301). Lee simply lacked the numbers. The combined effects of having dispatched Magruder to Malvern Hill, getting no help from Huger or Jackson at Glendale, and having Jackson fail to fix Federal reinforcements in front of him at White Oak Bridge left Lee hamstrung. In the end, he could claim a narrow tactical victory at Glendale, but he certainly failed in his plan to cut the Federal army in half. D. H. Hill pragmatically concluded, "It had been a gallant fight . . . but as an obstruction to the Federal retreat . . . amounted to nothing" (Wheeler 335).

The morning after the battle, when an officer told him that McClellan might escape, Lee replied with an uncharacteristically hot temper, "Yes, he will get away because I cannot have my orders carried out" (Sears, *Gates* 277–307). Freeman sympathizes with Lee, writing that at Glendale, "the story of the day should have been one of a driving march and of a straining effort by every column to reach its assigned place at the earliest moment. Instead, the events of June 30 fall into a succession of delays, of groping marches, of separate decisions" (*Lee's Lieutenants* 1:566–67). At Glendale, Lee's lieutenants failed him.

However, it must also be noted that poor staff work made this failure easy to come by. There were no arrangements made for continuing communications as the separate columns advanced and no procedure for the individual commanders to keep Lee informed. The result was that Lee had little situational awareness, and each subordinate commander operated too independently and without coordination (B. Alexander 131). Much of this could have been alleviated by a staff organization designed to help the commander control the battle.

MALVERN HILL

Facing Holmes and Magruder at Malvern Hill was a considerable force, including most of the Army of Potomac's artillery reserve. Seven batteries, totaling thirty-six guns, were deployed on Malvern Cliffs overlooking River Road to the west. The army had also reached the protection of the Federal fleet on the James, and the ships' guns would be used to great effect against the Confederates (Freeman, *Lee's Lieutenants* 1:583). Added to this array, Fitz John Porter also had nearly two divisions from the Fifth Corps in the area, and they would soon be joined by all the retreating Federal forces. Malvern Hill would be a formidable position.

On July 1, McClellan made a quick inspection of the Malvern Hill line, but by 9:15 A.M., he was back onboard the *Galena* heading toward Harrison's Landing, the army's ultimate destination. Again, he made no provisions for a chain of command in his absence. By virtue of being in command at the point of attack, Fitz John Porter would become the de facto commander of the Federal forces at Malvern Hill.

Malvern Hill was not so much a hill as an elevated, open plateau. The area to its northeast, the direction from which the Confederates were coming, was entirely open. Porter's artillery would have what Freeman describes as "a perfect field of fire for 300 or 400 yards" (*Lee's Lieutenants* 1:593). Martin Schenck describes the Federal position: "The lines of guns were banked three deep on the slopes and were arranged in a convex semicircle covering all approaches that were open to Lee, right, front, and left. Behind them were the James and the Federal supply base, so that ammunition was plentiful and there was no danger from a rear attack. The arrangement has been characterized as an artilleryman's dream" (94).

But the strength of Malvern Hill was also its potential weakness. The terrain led Porter to deploy his artillery and infantry along a narrow front just twelve hundred yards wide. A determined Confederate assault, as at Gaines' Mill, could penetrate the line before reinforcements could plug the gap. This was indeed a possibility, as Malvern Hill represented the first time during the Seven Days that the Army of Northern Virginia was united on a single field.

Lee's plan was for Jackson to lead the Confederate advance from Glendale along Quaker Road. Magruder would follow. Two of Huger's brigades under Lewis Armistead and Rans Wright were up already and would move west on a road leading off of Long Bridge Road into the Carter Farm. Huger's other two brigades and Huger himself would arrive later. Longstreet would personally remain with Lee during the battle, but Longstreet's command as well as that of A. P. Hill would constitute the reserve at Carter Farm.

Freeman observes, "These orders seemed simple enough, but from the time they were put in execution co-ordinated effort virtually ended. Although the divisional leaders were not apart at any hour, most of them lost touch" (*Lee's Lieutenants* 1:590). For this reason, Freeman titles this chapter of *Lee's Lieutenants,* "Malvern Hill: A Tragedy of Staff" (1:588). Again, the staff would do little to help the commander control events to ensure that they unfolded in accordance with his intent.

This plan first went awry as Magruder ended up on the wrong road. Several roads in the area were known as Quaker Road, and Magruder wound up on one that led him obliquely away from the day's battlefield. By the time he got reoriented in the right direction, three hours had been wasted. Thus, when Jackson arrived at Malvern Hill and took his assigned position on the left, he was alone. The plan had been for Magruder to come up on the right, but he was now nowhere in sight. Lee's plan was already unraveling.

While the Army of Northern Virginia struggled to reach the objective, Lee and Longstreet reconnoitered Malvern Hill. They found positions on the left and right that would each accommodate batteries of up to sixty guns that could punish the Federal force in a crossfire. Unfortunately, events would not work out that way.

Magruder's marching error can at least be partially explained by the lack of quality maps that continued to plague both armies. The next mistake that would hamper Lee was the poor manner in which his intention for this artillery barrage was transmitted.

Lee's chief of staff, Colonel Robert Chilton, penned an order stating, "Batteries have been established to act upon the enemy's line. If it is broken as is probable, Armistead, who can witness effect of the fire, has been ordered to charge with a yell. Do the same." This order left the initiation of an army attack entirely to the judgment of one brigade commander, Lewis Armistead, who was commanding in his first battle. Moreover, the only signal for a simultaneous charge of fifteen brigades was the Rebel yell of a single brigade, a signal very likely to be lost or misconstrued in the din of battle. Finally, Chilton did not annotate the order with a time so that it could later be placed in context with other orders and developing events.

Chilton, however, should not shoulder all the blame. Lee apparently did not help write or even review the order. Such an omission by the commander with regard to such an important instruction invites disaster. Lee's subordinates had hardly demonstrated throughout the Seven Days that they were able to carry out his instructions without a large degree of supervision. Malvern Hill would prove to be more of the same.

The next Confederate failure at Malvern Hill was the inability to establish the artillery crossfire. On the left, Jackson was down to just ten batteries, and they were widely scattered. His other seven, from D. H. Hill's division, were still in the rear being resupplied after firing all their ammunition on June 30. To make matters worse, Stapleton Crutchfield, Jackson's artillery chief, was sick and in the rear.

A potential solution lay in Brigadier General William Pendleton's artillery reserve, but he would show little initiative on July 1. Consequently, of Pendleton's fourteen batteries of reserve artillery, just one would be employed. One of Pendleton's battalion commanders would later complain of "the great superabundance of artillery and the scant use that was made of it" at Malvern Hill (Freeman, *Lee's Lieutenants*

1:618). In all, thirty-five Confederate batteries that were available for action around Malvern Hill would fail to fire a round.

The blame, however, is not all Pendleton's. True, he showed little initiative, but equally true he received no orders or instructions (Cullen 156). This is yet another example of the poor staff work that plagued the Confederates during the Seven Days.

Another aspect working against the Confederate massing of artillery was the fact that at this point in the war, Confederate artillery was dispersed so that a battery was attached to each brigade of infantry. Porter Alexander notes that "this scattering of the commands made it impossible to mass our guns in effective numbers. For artillery loses effect if scattered" (E. Alexander 104).

Against these odds, the duel began. The batteries on the left opened fire before those on the right, allowing eight Federal batteries of thirty-seven guns to concentrate against the three Confederate batteries of sixteen guns. The Confederate left battery was soundly defeated.

The right battery was also having its problems. Of Huger's six batteries, only two got to the front and into action. Of Magruder's sixteen, only two opened fire. Of Pendleton's fourteen reserve batteries, only one engaged. Seeing the situation, A. P. Hill sent his best battery forward, giving the Confederate right a total of just six batteries. These were no match for the Federals.

The infantry battle began when Federal skirmishers crept to within rifle range of the Confederate artillery. Armistead felt the need to drive them back, which he did easily. In the process, however, he drew a more significant enemy fire, and his men found themselves pinned down in the cover of a shallow ravine. The Confederate advance was reported to Lee as a success in spite of the fact that Armistead's men were now trapped by Federal fire. If the Federal response to Armistead's feeble attack had been planned as a ruse, it could not have had better effect (B. Alexander 126).

At about this time, Magruder finally arrived on the scene with his men filtering in behind him. Magruder's aide, Captain A. G. Dickinson, reported to Lee and brought back instructions that "General Lee expects you to advance rapidly. He says it is reported the enemy is getting off. Press forward your whole line and follow up Armistead's success."

Magruder naturally interpreted this to represent an order for an immediate attack, but Lee apparently meant otherwise. That night, Lee asked Magruder, "General Magruder, why did you attack?" seeming to indicate that Lee expected Magruder to prepare as necessary and attack when the situation was right (Sears, *Gates* 335). If so, Lee certainly did not transmit this intention to Magruder at the time.

Part of the problem was that as Magruder arrived on the battlefield, he was handed the order Chilton had written three hours earlier outlining the original plan of attack. With the failure of the artillery bombardment, this order was now quite outdated, but because it lacked a time annotation, Magruder had no means of placing it in chronological context. Receiving Dickson's message fresh on the heels of Chilton's reinforced the idea that Lee wanted an immediate attack. Furthermore, Magruder was still smarting from Lee's rebuke of his lack of aggressiveness at Savage's Station and was not about to be found wanting in this department again. Magruder would attack immediately.

With this decision, Magruder committed himself to a piecemeal attack with forces as they became available. To do so, he first used Huger's troops, the brigades of Wright and Billy Mahone and half of Armistead's. With these five thousand men and little artillery behind them, Magruder attacked at 5:30 P.M. The forces quickly became separated, and the attack became uncoordinated. This would be the story of the day for the Confederates. D. H. Hill would later throw his five brigades of eighty-two hundred men into the fight in what Stephen Sears describes as not "one unified sweep forward to overwhelm the Federal line, [but] five separate attacks" (Sears, *Gates* 326). At Gaines' Mill, Lee had won by finally getting all his forces assembled and under control for a massed attack that overwhelmed the Federals. At Malvern Hill, he would not be able to repeat this formula.

The end result for the Army of Northern Virginia was 869 dead, 4,241 wounded, and 540 missing. The Federals lost half that many: 314 killed, 1,875 wounded, and 818 missing. Summing up the battle, D. H. Hill wrote that the Confederates "did not move together, and were beaten in detail. As each brigade emerged from the woods, from fifty to one hundred guns opened up on it. . . . It was not war—it was murder." Furthermore, Hill estimated that more than half of the Confederate losses had resulted from artillery (McWhiney and Jamieson 113).

Indeed, the decisive factor at Malvern Hill had been the Federal artillery. Grady McWhiney and Perry Jamieson describe Malvern Hill as "perhaps the most famous use of artillery on the defensive during the

Civil War." In addition to Hill, several other Confederates noted the effect of the Federal artillery. Stephen Lee, himself a Confederate artilleryman, credited the Federals at Malvern Hill with having produced "the most terrible artillery fire during the war." Heros von Borcke, a former Prussian cavalry officer, wrote that the "effect was more disastrous than had been before produced by artillery" (McWhiney and Jamieson 112–13). The Federal artillery clearly carried the day at Malvern Hill.

At 9:30 that evening, Porter sent word to McClellan that "against immense odds, we have driven the enemy beyond the battle field and the firing ended at 8:30." Moreover, Porter, generally nearly as cautious as McClellan himself, advised that, if resupplied with food and ammunition, "we will hold our own and advance if you wish." In spite of Porter's success at Malvern Hill, McClellan would entertain no such thoughts (Sears, *Gates* 291–336).

If "the commander's will is the constant element that propels the force through the shock and friction of battle" (FM 3-0 5-75), the Army of the Potomac would certainly benefit from no such fortitude on the part of McClellan. The retreat of the Army of the Potomac would continue. As James Murfin summarizes, "Despite superior artillery positions, advantages in numbers, and generals who were willing to fight and fought well, McClellan did not once consider an offensive. Despite clear-cut advantages throughout the Seven Days, McClellan was determined he would not be run over by an army that [he imagined] outnumbered him two to one" (56).

EVELYNTON HEIGHTS

The Confederates had one last opportunity to destroy the Federal army on the Peninsula. On June 30, Lee ordered Jeb Stuart to rejoin the Army of Northern Virginia and "cooperate" with Jackson. Pursuant to these orders, Stuart sent Captain John Pelham with one twelve-pound howitzer and a squad of the First Virginia Cavalry ahead to reconnoiter McClellan's position.

Pelham discovered a glaring weakness in the Federal position at Harrison's Landing. Evelynton Heights, a slightly elevated plateau that completely commanded Harrison's Landing, was inexplicably unoccupied. Pelham quickly notified Stuart of the situation but wisely refrained from taking any further action to avoid alerting the Federals of their vulnerability.

Stuart, however, would not be so circumspect. He impetuously seized the heights and began firing on the entire Army of the Potomac with Pelham's lone howitzer. The Federal infantry quickly advanced against him and forced Stuart to withdraw. Thus was lost an excellent opportunity for a larger force to exploit the Evelynton Heights advantage (Cullen 164–65).

Longstreet would later want to assault the heights even though surprise had been lost, but Jackson argued that his troops were in no condition for another frontal assault. When Lee surveyed the position, he agreed with Jackson. The Federal position was too strong.

The Army of Northern Virginia had accomplished its strategic objective of turning back the Federal threat to Richmond, but Lee's army had also reached its point of culmination. In fact, the campaign had taken a greater toll on the victorious Confederates, who suffered 20,141 total casualties (about 22 percent of the force), than on the defeated Federals, who suffered 15,849 (about 18 percent of the force). Martin Schenck concludes that "both armies were damaged but the one hurt worse was the defender and not the invader" (96). By July 8, all of the Army of Northern Virginia, with the exception of cavalry outposts and pickets, was back in the camps around Richmond for a much needed rest and reorganization (Cullen 165).

One of the key factors that convinced Lee of the wisdom of this course was the powerful presence of the Federal navy on the James. On July 6, he wrote to President Davis, "The great obstacle to operations here is the presence of the enemy's gunboats, which protect our approaches to him, and should we even force him from his positions on his land front, would prevent us from reaping any of the fruits of victory and expose our men to great destruction" (Schenck 99).

Throughout the campaign, McClellan and Goldsborough had had one problem after another in affecting joint cooperation, but the two services certainly worked well together during the withdrawal. Bern Anderson goes so far as to say that "the Union Navy saved McClellan's army from probable destruction on the banks of the James" (84). Given the exhausted state of the Confederate army, this may be a bit of an overstatement, but this fear for the survivability of the army seems to have been what was needed to solve McClellan and Goldsborough's previous problems with effective communication and shared purpose.

CHAPTER 5

SO WHAT?

The great nineteenth-century military theorist Antoine-Henri de Jomini felt that there were three kinds of military history: a pure version that recounts in minute and pedantic terms all aspects of a given battle, an analytical version that uses the battle to examine general principles, and a political-military version that examines war in its broadest context through association with political, social, and economic factors (1–21). The purpose of this particular study is best captured by Jomini's analytical version. By analyzing the relationship between events and principles and by studying the campaign as a part of a broader context of warfare as an evolutionary process, certain truths can be gleaned to advance the understanding of the art of waging war. With this in mind, our attention will now focus on critical reviews of McClellan's offensive, Lee's defense and counteroffensive, Jackson's performance during the Seven Days, and McClellan's withdrawal.

A CRITICAL REVIEW OF McCLELLAN'S OFFENSIVE

Few can fault McClellan as an administrator, builder of an army, or planner. Instead, his failings lay in the execution. In Clifford Dowdey's assessment, "McClellan's strategy for taking Richmond was flawless. It was the only imaginative plan ever used against the city and the only plan of any kind that nearly succeeded. What he might have done had he been allowed to follow this plan for the conquest of Richmond can never be known. He was thwarted by a strange triumvirate—Lee, Jackson, and Lincoln" (*Land* 164). Dowdey, however, omits one important factor. To this triumvirate, one must add McClellan as well.

Military theorists and strategists have long used the principles of war as a means of analyzing military operations. These are objective, offensive, mass, economy of force, maneuver, unity of command, security, surprise, and simplicity (FM 3-0 4-12). These principles provide a convenient framework for analyzing McClellan's actions on the Peninsula.

OFFENSIVE AND MANEUVER

The offensive principle of war tells us that "offensive action is key to achieving decisive results" (FM 3-0 4-13). At the tactical level, this notion was lost on McClellan in his preference for the methodical and low-risk siege method of warfare. However, at the operational level, albeit with some prodding from Lincoln, McClellan eventually assumed the offensive by launching the Peninsula Campaign.

The problem with achieving the principle of offensive, especially given the nature of the Civil War battlefield, is to overcome the inherent advantages of the defender. Herman Hattaway and Archer Jones note that "If both the attacking and defending combatants desire a battle then one will occur, usually where and when the defending force chooses and with the defender enjoying the advantage. If the defender does not care to fight, he can retreat or intrench so that the attacker has no hope of success except by employing superior numbers, if he has them, to threaten the defender's flanks and force a retreat, a slow process" (82).

This problem is overcome by the application of the maneuver principle of war. Maneuver is placing "the enemy in a disadvantageous position through the flexible application of combat power" (FM 3-0 4-14). The question then became how McClellan would maneuver his force to overcome the Confederate positional advantage of being between Washington (where McClellan was) and Richmond (where he wanted to go). Hattaway and Jones conclude that for McClellan, "the classic solution to this problem was the turning movement" (82).

The turning movement "uses freedom of maneuver to create a decisive point where the enemy is unprepared." It is executed by bypassing the enemy's defenses and securing key terrain and lines of communication deep in the enemy's rear. The defender is then forced to abandon his prepared defenses and fight in the open to meet this unexpected threat. In this way, the attacker avoids the defender's concentrated fires

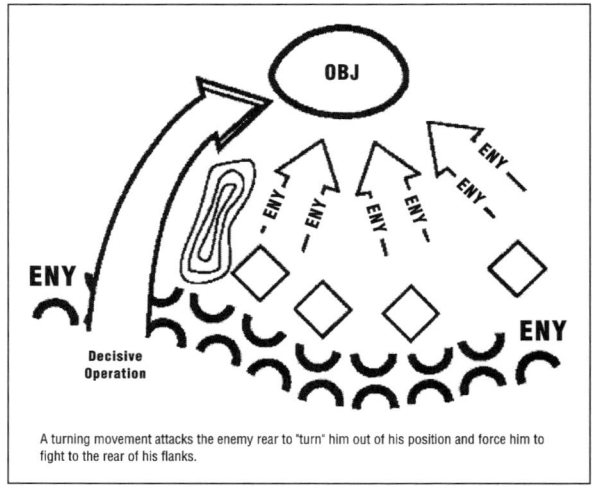

A turning movement attacks the enemy rear to "turn" him out of his position and force him to fight to the rear of his flanks.

Typical turning movement (FM 3-0 7-12)

that would be faced in a frontal attack (FM 3-0 7-12). The amphibious assault is a common form of turning movement.

McClellan lost much of the effects of a turning movement when Johnston's withdrawal led to the abandonment of the Urbanna Plan. Even thereafter, however, McClellan still had plans to land either north of Gloucester Point or below Yorktown and turn the Confederate defenses. McClellan states that the slowness of acquiring transportation forced him to change his plan and land instead at Fort Monroe (168). The result was that the actual attack looked much more like a frontal assault than a turning movement; thus, B. H. Liddell Hart characterizes the campaign as more "of a shorter direct approach to Richmond . . . than an indirect approach in the true sense" (126).

Even after the Urbanna Plan became obsolete, McClellan had at least three other options for his amphibious movement. The first would have been to land either north of Gloucester or south of Yorktown and turn the Confederate defenses, as McClellan himself mentions. The second was to land at Fort Monroe, as McClellan ultimately did. A third, however, would have been to land at multiple locations. Indeed, Joint Pub 3-02.1, *Joint Doctrine for Landing Force Operations,* states that "the [amphibious] assault is made on the widest front with the greatest depth possible, consistent with the capability of the landing force to accomplish its mission ashore. . . . Separation of units is accomplished through the use of multiple, separated landing beaches and landing zones within the landing area" (V-11).

By landing simultaneously at Fort Monroe and at a point south of Yorktown, McClellan could have fixed Magruder to his front, gotten behind his first line of defense, and beat the bulk of Magruder's forces to his Yorktown-Warwick line. Likewise, McClellan could have landed at Fort Monroe with one force and just below Gloucester Point with another. With the Fort Monroe force, he could have fixed Magruder along his Yorktown-Warwick line, and with the other he could have bypassed Gloucester and moved along the east bank of the York toward Richmond or reduced Gloucester and, with pressure on Yorktown, opened the York to Goldsborough's gunboats. Any of these options would have given Magruder more than one force to contend with.

Another option would have been to land forces both on the Peninsula and in North Carolina to reinforce Burnside. Such a course of action would have helped mitigate Lee's ability to concentrate his forces by causing him to respond to multiple threats. Indeed, Joint Pub 3-02.1 states, "The cumulative shock effect resulting from dispersed, coordinated assault landings on beaches and movements of assault elements inland causes disruption of enemy resistance over an extended area in both width and depth" (V-12). Lee suffered no such disruption.

The overall problem was that the principle of offensive requires an audacity that was beyond McClellan's capacity. McClellan fought not to lose rather than to win. George Meade summed up McClellan's failure on the Peninsula by saying, "McClellan was always waiting to have everything just as he wanted before he would attack, and before he could get things arranged as he wanted them, the enemy pounced on him and thwarted all his plans. . . . Such a general will never command success, though he may avoid disaster" (Murfin 57). McClellan considered escaping without his army being destroyed a great Federal victory. Few would agree.

SURPRISE

Part of the reason for the lack of disruption to Lee was that McClellan did not adhere to the surprise principle of war, the goal of which is to "strike the enemy at a time or place or in a manner for which he is unprepared" (FM 3-0 4-14). Joint Pub 3-02, *Joint Doctrine for Amphibious Operations,* reminds us that "the amphibious operation exploits the element of surprise" (I-1). McClellan made no efforts to do so. Even going back to the original Urbanna Plan, Archer Jones notes that "McClellan had no plans to distract the enemy and so none for attaining surprise. Without surprise, he could not expect to trap the rebel army and could only aim at the more limited but more realistic objective of forcing it back" (60). A more imaginative landing plan that included some deceptive measures would certainly have helped McClellan achieve surprise.

The other factor hindering McClellan in achieving surprise was his predictability. McClellan and his deliberate and cautious approach to warfare were well known to his Confederate counterparts. Lee was especially aware of this characteristic and exploited it to great effect. It was the perfect vulnerability that Lee could exploit with his own opposite characteristic of audacity.

SECURITY

Somewhat of the inverse of surprise is the security principle of war. The purpose of security is to "never permit the enemy to acquire unexpected advantage" (FM 3-0 4-14). McClellan was terrible at this. Stephen Sears observes that McClellan "was invariably brought up short by the unexpected" (*McClellan* 175).

Part of this problem was the fact that McClellan lacked a cavalry organization capable of providing the quality intelligence that Stuart provided for Lee. Instead, McClellan received only Pinkerton's inaccurate and highly exaggerated reports.

However, McClellan must also take the blame for his inflexibility, which caused him to fail to adapt to changing situations and deviate from his original plans. Strategic leaders must operate in what today's army calls the VUCA environment—volatile, uncertain, complex, and ambiguous. They "fight complexity by encompassing it," and they "must be more complex than the situations they face" (FM 22-100 7-9).

McClellan's talents lay in the more orderly world of administrating and organizing an army. He could not adjust to the VUCA environment in which he would have to fight that army. Examples are his fixation with an amphibious movement even after Johnston's withdrawal from the

Centreville-Manassas line had negated many of its advantages, McClellan's failure to foresee the need to plan a branch to exploit the Confederates' abandonment of Yorktown, and his failure to take advantage of the opening of the James. A strategic leader must be able "to expand [the] frame of reference to fit a situation rather than reducing a situation to fit . . . preconceptions" (FM 22-100 7-9). McClellan failed in this regard.

UNITY OF COMMAND

Throughout the Peninsula Campaign, McClellan failed to achieve unity of command. His difficulties in establishing such a relationship with Goldsborough can be partially explained by the underdeveloped stage of joint operations at the time. Nonetheless, McClellan can be faulted for failing to establish unity of effort through the personal relationship skills, effective communication, and shared purpose that Grant and Foote achieved at Fort Henry.

But McClellan's unity of command problems extended on land as well, and in this area he has no one to blame but himself. Throughout the Peninsula Campaign, McClellan was found to the rear of the battle, leaving the direction of events to his subordinate commanders, and throughout the Seven Days, McClellan failed to establish a chain of command in his absence. Consequently, the individual corps commanders' efforts were never coordinated or mutually supporting. "The will and personal presence of commanders provide the impetus for action" (FM 3-0 5-75). McClellan offered neither.

This condition was partially offset by the work of McClellan's staff, which clearly outperformed Lee's. An example of the Federal staff work is engineer John Barnard's identification of subsequent positions, notably at Malvern Hill, during the withdrawal (Cullen 151–52). Nonetheless, in the equation of command and control, the staff can assist the commander in the area of control, but the commander must command. McClellan frequently abdicated this responsibility on the Peninsula.

OBJECTIVE

Perhaps McClellan's greatest violation of the principles of war was his failure to adhere to the principle of objective. Under this principle, every military operation is directed toward a clearly defined, decisive, and attainable objective. This "normally includes aspects of the political dimension" (FM 3-0 4-12), and indeed, McClellan's most serious failing here lay primarily in the political rather than military context. McClellan simply could not reconcile, let alone subordinate, his objectives to President Lincoln's.

In *On War,* Carl von Clausewitz writes, "No major proposal required for war can be worked out in ignorance of political factors" (608), but McClellan's actions ignored this requirement. Hattaway and Jones observe that McClellan "seemed unable to grasp the political realities faced by an administration which sought to draw hostile elements into full support of the war for the Union" (98–99).

This represents one of McClellan's greatest shortcomings as Lincoln's subordinate—the failure to understand the president's developing grand strategy. Lincoln was considering options to expand the parameters of the war by drawing on a wide range of military, political, and economic weapons at his disposal. McClellan, as a result of both his political beliefs and his deliberate nature, favored a much more restrictive and limited war. As Joseph Glatthaar observes, "Their opinions gravitated toward opposite poles" (81).

The lesson for today's military leadership is the need to ensure that the National Military Strategy is synchronized with the National Security Strategy and the Defense Strategy. Through the National Security Strategy, the president articulates the national interests and his approach to advancing them. This strategy incorporates all of the elements of national power, not just the military. Through the National Military Strategy, the chairman of the Joint Chiefs of Staff gives advice on how the military will contribute to fulfilling its responsibilities in the National Security Strategy and the Defense Strategy. The military's efforts must be synchronized and integrated with the other instruments of national power. This is how the principle of war of objective is achieved at the highest level. The military cannot go its own way. Current army leadership doctrine requires that "strategic leaders . . . draw on their conceptual skills to comprehend national, national security, and theater strategies [and] operate in the strategic and theater contexts" (FM 22-100 7-7). McClellan simply could not understand this. His conflict with the Lincoln administration over emancipation is a good example.

It is also the military's responsibility to keep the president informed. In this regard, McClellan made no genuine attempt to assuage Lincoln's fears for the safety of Washington. Glatthaar concludes, "Since McClellan did not fear a Confederate advance on Washington, in his view no menace existed. It never dawned on him that the commander in chief might view matters differently" (73). The sparse information that McClellan shared with the president was cryptic and unintelligible. Hattaway and Jones note that "though [McClellan] explained the power of the defense and the strategy of the turning movement, the sophistication of these belatedly explained ideas made them difficult for the President to grasp" (98–99).

Today's operations are infinitely more complicated than those during the Civil War. Whereas McClellan had to coordinate objectives primarily with just the president and the navy, the military now finds itself cooperating with a host of coalition partners, nongovernment organizations, and other governmental agencies in an effort to achieve unified action (FM 3-0 2-1). The difficulties in such an environment include reconciling sometimes competing agendas and building consensus. This is a major portion of the interagency dynamic.

McClellan made no effort to understand the president's agenda. He cared only about his own. As General Douglas MacArthur would learn nearly a century later, such an attitude will not work. One cannot help but think that if McClellan had sat down with Lincoln and fully discussed the issue, the two men could have reached a more productive arrangement. The same is true of McClellan's cooperation with Goldsborough. In today's less clear-cut situations, military leaders must apply the same lesson to their dealings with other agencies, such as the Departments of Justice, Transportation, Homeland Security, and the Treasury, that may be involved in the operation and have their own unique concerns and agendas.

A large part of McClellan's political failure was his unabashed contempt for Lincoln and failure to subordinate himself to civilian authority. T. Harry Williams contends that McClellan assumed "he was big enough to treat with the government as an equal party" (*Lincoln* 109). This attitude is perhaps best understood by a sampling of McClellan's comments on the government's leadership. He considered Lincoln to be "an idiot" and "nothing more than a well-meaning baboon." Secretary of State William Seward was a "meddling, officious, incompetent puppy."

Secretary of the Navy Gideon Welles was "weaker than the most garru-lous woman you were ever annoyed by." Attorney General Edward Bates was "an old fool." General in Chief Winfield Scott was "a perfect imbe-cile." Indeed, Glatthaar concludes that "Anyone in the administration who crossed McClellan's path and did not yield wholeheartedly to his program came in for such epithets" (61). Sears agrees, writing that McClellan believed that "anyone who did not support him in every aspect and without exception must be against him" (*McClellan* 149).

A scenario completely unimaginable today illustrates the depth of McClellan's contempt and insubordination. One evening, Lincoln, Seward, and presidential secretary John Hay dropped by McClellan's headquarters, only to learn that McClellan was at a wedding and would return shortly. McClellan returned in about an hour and, upon being notified of his visitors, went upstairs anyway. After waiting another thirty minutes, Lincoln dispatched a servant to remind McClellan that the pres-ident was still downstairs. Moments later, the servant returned with a message that McClellan had gone to bed for the evening (Glatthaar 64). That Lincoln could maturely and selflessly overlook such a snub in defer-ence to the greater emergencies of the time speaks as much about the president's character as McClellan's egotistical behavior does about his.

McClellan also failed to exercise the principle of objective at the tacti-cal level. Commanders select objectives based on the concept of the deci-sive point. The decisive point "is a geographic place, specific key event, or enabling system that allows commanders to gain a marked advantage over an enemy and greatly influence the outcome of an attack" (FM 3-0 5-7). McClellan knew that Yorktown met this definition when he described it as "the most important point—there the knot to be cut" (Sears, *McClellan* 173). Nonetheless, McClellan did nothing to ensure suc-cess at the decisive point beyond his token coordination with the navy.

In sum, it appears that most of McClellan's problems, be they mili-tary or political, stemmed from his character and personality. Peter Parish provides an excellent analysis:

[McClellan's] defects were his virtues carried to excess. His confidence became arrogance, or mere vanity. His thoroughness led him into over-caution, procrasti-nation, and a perfectionism to which the hard and swiftly-changing realities of war could seldom conform. His flair for organization and planning betrayed him into inflexibility, a tendency to assume that the facts must inevitably fit his plans, and

an inability to improvise or seize an unexpected opportunity. His military profes-
sionalism degenerated into a narrow exclusiveness, an ill-disguised contempt for
his political masters, and a willful blindness to the non-military considerations
which virtually affected grand strategy. His sense of mission caused him to see all
who crossed him as conspirators and himself as a martyr. He had most of the talents, except the ability to put them to the greatest use. (192)

A CRITICAL REVIEW OF LEE'S DEFENSE AND COUNTEROFFENSIVE

Just as the principles of war provide a convenient frame-
work for military analysis, a similar purpose is served by the facets of the
operational art: synergy, simultaneity and depth, anticipation, balance,
leverage, timing and tempo, operational reach and approach, forces and
functions, arranging operations, centers of gravity, direct versus indirect, decisive points, culmination, and termination (Joint Pub 3-0 III-
10). Chief among these factors that are apparent in Lee's conduct on the
Peninsula are simultaneity and depth, anticipation, balance, timing and
tempo, arranging operations, and direct versus indirect. Culmination is
also critical, but it will be discussed separately in the context of its
impact on Jackson's efforts.

SIMULTANEITY AND DEPTH

The intent of simultaneity and depth is to create competing and simultaneous demands throughout the battle area within the
enemy's decision-making cycle (Joint Pub 3-0 III-11-12). In his role as
military adviser at the outset of the Peninsula Campaign, Lee developed
simultaneity and depth in the Confederate response to McClellan's
attack by involving Jackson's army in the Valley. If Jackson could create
a threat to Washington, Lee figured that forces McClellan had intended
to use against Richmond would be diverted. Lee was right, and
McDowell's First Corps was denied to McClellan because of the threat
posed by Jackson.

ANTICIPATION

Through anticipation, commanders "remain alert for the unexpected and for opportunities to exploit the situation" (Joint Pub 3-0 III-12). Critical to this capability is adequate intelligence of the enemy, and in this area Lee had a marked advantage over McClellan. Lee recognized the importance of intelligence and greatly benefited from the collection capability afforded him by Stuart's cavalry. McClellan had no such capability and was in fact hindered by his intelligence service's efforts.

Collection, however, is but one step in the intelligence cycle. What truly defined Lee's superior use of intelligence was his ability to incorporate it into his planning. An excellent example of this is his plan to turn McClellan's right flank at Mechanicsville that resulted from Stuart's report of that Federal vulnerability.

Commanders must be able to filter through tremendous amounts of information using the skills of analysis and synthesis. Analysis breaks a problem down into its component parts. Synthesis assembles complex and disorganized data into a solution (FM 22-100 6-7). Lee's ability to analyze Magruder's initial panicky reports and synthesize them into the need for the reconcentration of forces shows Lee's mastery of this skill. James Murfin seconds this assessment of Lee's analytical prowess, concluding that on the Peninsula, "in nearly every situation, Lee would analyze [McClellan] correctly" (53).

Lee also used what is now called intelligence preparation of the battlefield to help him identify likely enemy courses of action. He affected this in the broad sense with his understanding of McClellan's cautious nature and then specifically focused this understanding with reconnaissance. An example of this is when Lee realized after Gaines' Mill that McClellan was withdrawing across the Chickahominy. Lee identified three possible intentions of McClellan's movement and then sent out reconnaissance missions under Ewell and Stuart to confirm or deny these possible enemy courses of action. From these reconnaissances, Lee learned that the Federals were in full-scale retreat, and he then developed a plan to intercept them.

But one area in which Lee did not demonstrate effective anticipation was that, after ascertaining that McClellan was in retreat, Lee failed to block the Federal general's path. After determining that McClellan was heading for Harrison's Landing, Lee could have used Stuart's cavalry to reconnoiter a suitable position, such as Malvern Hill, from which to cut off the retreat. Then Lee could have deployed a force to beat McClellan to this position. But instead of cutting McClellan off, Lee ended up striking the Federals on the march in what turned out to be what Bevin Alexander describes as only "glancing blows on the flank and rear." Alexander correctly assesses that the Federal army "could be destroyed only by blocking its passage and attacking it front and rear" (116–17). More effective anticipation would have afforded Lee this opportunity.

BALANCE

Balance is "the maintenance of the force, its capabilities, and its operations in such a manner as to contribute to freedom of action and responsiveness" (Joint Pub 3-0 III-13). One excellent example of Lee's achievement of balance is his response to the dire initial reports of McClellan's amphibious move. In fact, one of the purposes of an amphibious movement is to throw the enemy off balance. To this end, Joint Pub 3-02, *Joint Doctrine for Amphibious Operations,* notes, "The threat of an amphibious landing can induce enemies to divert forces, fix defensive positions, divert major resources to coastal defense, or disperse forces. Such a threat may result in the enemy making expensive and wasteful efforts in attempting to defend the coastline" (I-1). Lee resisted the temptation to make such wasteful efforts. In spite of panicked reports from his subordinates, he remained calm. Instead of dispersing forces to meet unconfirmed threats, Lee concentrated his forces in such a way that gave him the flexibility to meet threats as they developed. As President Davis's military adviser, Lee correctly maintained a strategic rather than tactical view of the situation. This vision and the resultant balanced response allowed the Confederates to use their scarce resources to the greatest effect.

TIMING AND TEMPO

Through timing and tempo, commanders "conduct operations at a tempo and point in time that best exploits friendly capabilities and inhibits the enemy" (Joint Pub 3-0 III-15). Thus, at the beginning of the Peninsula Campaign, Confederate forces fought a delay and traded space for time while Lee affected his reconcentration of forces. McClellan helped the Confederates set the timing and tempo of this stage of the campaign by resorting to a time-consuming siege.

When the initiative shifted to the Confederates after Seven Pines, Lee kept up relentless pressure and a fast pace of operations during the Seven Days. McClellan now wanted to trade space for time as he withdrew to the protection of the navy's gunboats, and Lee was determined to deny the Federal general this luxury. By adjusting the timing and tempo, Lee took control of the campaign.

But Lee would control the timing and tempo of the fighting well beyond the Peninsula Campaign. His conduct during the Seven Days marked the beginning of a new strategy for the Confederates: the offensive-defensive (Stiles 33). Through it, Lee would combine economy of force, audacity, and maneuver to fight the Federal army throughout the rest of the Civil War.

ARRANGING OPERATIONS

One facet of the operational art that was a continual source of frustration for Lee during the Seven Days was that of arranging operations. Beginning with Jackson's late arrival at Mechanicsville, Lee was never able to synchronize his attacks to bring his forces to bear at the decisive place and time. Consequently, he was never able to adequately mass his forces or concentrate their effects. Lamenting these missed opportunities, Lee wrote, "Under ordinary circumstances the Federal army should have been destroyed." Russell Weigley concludes that "what [Lee] meant by 'ordinary circumstances' was a professional command and staff which would not have been guilty of the lapses he suffered during the campaign" (108). Freeman agrees, blaming many of the difficulties on what he calls "the worst imaginable staff work" (*Lee's Lieutenants* 1:604). Archer Jones is a little more understanding, assessing that "Lee and his staff, in their first campaign together, showed their inexperience" (70). After the Peninsula Campaign, Lee took the measure of his army and adjusted his tactics. He would not repeat attempts at complicated maneuvers such as the converging movements at Mechanicsville and

Glendale (Sears, *Gates* 344). A key lesson for the commander is that the operation must be simple enough for his forces to execute.

DIRECT VERSUS INDIRECT

Especially in the case of a numerically inferior force such as Lee's, it is often necessary to attack the enemy by an indirect rather than a direct approach. There are numerous examples of Lee's recognition of this requirement. The first is his strategic use of Jackson in the Valley indirectly to affect developments on the Peninsula. Operationally, the indirect approach often involves attacking enemy vulnerable points such as flanks and lines of communication (Joint Pub 3-0 III-21). Through turning movements and envelopments, Lee avoided the strength of McClellan's forces, and by threatening McClellan's line of communication with White House on the York River, Lee forced McClellan to withdraw.

A CRITICAL REVIEW OF JACKSON DURING THE SEVEN DAYS

After the brilliance of the Valley Campaign, Jackson's difficulties on the Peninsula have been the subject of much historical scrutiny and analysis. Various scholars have attempted partially to account for his poor performance: Jackson delayed at Grapevine Bridge because he wanted to keep his troops dry (Longstreet 131) or because he was afraid of Federal artillery (B. Alexander 120). He failed to march on Glendale because he had no orders to do so (Henderson 2:56–57; J. Robertson 497). He was unfamiliar with his surroundings and the geography of the country (Freeman, *Lee's Lieutenants* 1:497; J. Robertson 504; Hattaway and Jones 194). He did not understand his role in the time criticality of Lee's plan (Sears, *Gates* 197). He was brilliant when operating independently but chafed when compelled to closely cooperate with others (Vandiver 329). His tactical skill was not yet on par with his strategic prowess (Jones 70). His religious intensity caused him to put the battle in God's hands (E. Alexander 97–98). While some elements of all these explanations may be true, the basic reason for Jackson's lackluster showing on the Peninsula is that he was exhausted.

Those operations that follow up on an initial success and seek to extend destruction or completely destroy a fleeing enemy are called exploitations and pursuits. Such operations were the focus of the Confederate efforts during the Seven Days. Even with the technological advances of today's military, it is still recognized that "exploitations and pursuits test the audacity and endurance of soldiers and leaders alike. After an attack, soldiers are tired and units have suffered personnel and materiel losses. As an exploitation or pursuit unfolds, [lines of communication] extend and commanders risk culmination. Commanders and units must exert extraordinary physical and mental effort to sustain momentum, transition to other operations, and translate tactical success into operational or strategic victory" (FM 3-0 7-23).

One factor that makes successful exploitations and pursuits difficult is an army's tendency to reach its culmination, "the point in time and space where the attacker's effective combat power no longer exceeds that of the defender's or the attacker's momentum is no longer sustainable, or both" (FM 3-0 5-9). Jackson may indeed have reached his personal point of culmination before arriving on the Peninsula. On June 7, before the difficult battle at Port Republic, Jackson's aide, Sandie Pendleton, wrote in a letter to his mother that, due to physical and mental exertion, "General Jackson is completely broken down" (Tanner 360).

Freeman describes "physical exhaustion and the resulting benumbment of a mind that depended much on sleep . . . as the basic explanation of Jackson's inability to meet the demands of the campaign" (*Lee's Lieutenants* 1:659). Freeman's explanation for the slow march to Mechanicsville is that "after two long nights in the saddle, Jackson himself did not possess his normal energy and probably failed to realize that he lacked drive and grasp of the situation" (*Lee's Lieutenants* 1:498). Likewise, Freeman concludes that at Glendale, "Night found Jackson so weary, so confused, after almost twenty hours of profitless marching and waiting in unfamiliar country, that he was stupefied" (*Lee's Lieutenants* 1:579–80). The effect of this exhaustion was particularly acute given that Freeman describes Jackson as being "notoriously . . . dependent on abundant sleep" (*Lee's Lieutenants* 1:506; Freeman, *R. E. Lee* 2:578–82).

Examples of culmination abound in the Peninsula Campaign. On an individual level, leaders such as Magruder and McClellan suffered the effects of exhaustion and stress. On a collective level, the Confederate army, through casualties and repeated marches and attacks, eventually reached a point where it could no longer press the offensive. But perhaps the most poignant example of culmination is the personal culmination of Jackson. It stands as a caution to military leaders to include in their battle rhythm a suitable period of rest. To do otherwise risks the steady degradation of performance suffered by Jackson and the ultimate inability to function in a sustained manner.

A CRITICAL REVIEW OF McCLELLAN'S WITHDRAWAL

On June 28, 1862, McClellan began withdrawing his Army of the Potomac from White House to Harrison's Landing. Certain historians have questioned the wisdom of McClellan's decision, yet few can question its execution. In fact, while in many cases Civil War tactics appear archaic by today's standards, McClellan's withdrawal has stood the test of time. An analysis of McClellan's execution based on current military doctrine for retrograde operations indicates that McClellan's withdrawal was nearly flawless. Again, critics convincingly argue that McClellan's withdrawal was the result of unnecessary panic and pessimism, and there is ample evidence that he buckled under the stress and abdicated battlefield control to his subordinates while he squandered his time on rear-echelon staff duties (Glatthaar 78; Sears, *Gates* 281). Setting these legitimate criticisms aside, let us examine the Federal army's withdrawal as a military maneuver in and of itself.

Current doctrine states that withdrawals can be conducted for a variety of reasons (FM 3-0 8-6). The three that apply in the case of the Army of the Potomac are to preserve forces, to shorten lines of communication, and to facilitate repositioning forces. Of these, McClellan's principal concern was to preserve his forces.

McClellan's problems began with the heavy fighting around Mechanicsville on June 26, and his defeat at Gaines' Mill the following day served to exacerbate the situation. Only Jackson's uncharacteristic late arrival saved McClellan from decisive defeat. This narrow margin of escape caused McClellan to panic and greatly overexaggerate the size of his enemy. He fired off a series of anxious telegrams to Secretary of War Stanton with such comments as "Attacked by greatly superior numbers in all directions on this side . . . The odds have been immense . . . I may be forced to give up my position during the night" (Catton, *Terrible Swift Sword* 332). McClellan's overriding concern was obviously to avoid what he felt was inevitable defeat.

A secondary reason, perhaps based on rationalization and after-thought, was a desire to shorten the lines of communication. McClellan refers to this as his "change of base" (Dowdey, *Seven Days* 254). Stuart's ride around McClellan's army showed the vulnerability of the Federal supply line (Catton, *Terrible Swift Sword* 322). McClellan had to do some-thing about this. A move to Harrison's Landing would not only place him directly on the banks of the James River, within easy reach of the Fort Monroe stores, but would also allow any ensuing supply operations to be conducted under the protection of Goldsborough's gunboats.

A third advantage of the relocation would be to facilitate reposition-ing of forces. If McClellan's incessant requests for more troops were ever answered (indeed he would eventually receive five thousand garrison troops from Fort Monroe), Harrison's Landing would be a convenient linkup point. Thus, the withdrawal allowed for preservation of the force, resupply, and reinforcement (or perhaps evacuation if the situation worsened). Each of these factors is consistent with reasons prescribed by current doctrine for conducting a withdrawal.

After McClellan decided that a withdrawal was necessary, he had to task organize his command to accomplish the mission. Current doctrine states that a unit conducting a withdrawal under enemy pressure will organize into a security force and a main body. The security force's mis-sions are to conceal the main body's withdrawal and to deceive the enemy. If the enemy attacks during the withdrawal, the security force will provide covering fire (FM 3-0 8-6).

To this end, McClellan had both land and naval security forces. The naval security force consisted of Goldsborough and his gunboats. McClellan advised Goldsborough, "We have met a severe repulse today, having been attacked by greatly superior numbers, and I am obliged to fall back between the Chickahominy and the James River. I look to you to give me all the support you can in covering my flank as well as in giv-ing protection to my supplies afloat in the James River" (Catton, *Terrible Swift Sword* 333). To this point in the campaign, there had been one problem after another with cooperation between the army and the navy, but the two services worked well together during the withdrawal. Bern Anderson goes so far as to say that "the Union Navy saved McClellan's army from probable destruction on the banks of the James" (84). Indeed, it seems that this fear for the survivability of the army was required to solve McClellan and Goldsborough's previous problems had with effective communication and shared purpose. T. Harry Williams is also generous with his praise for the Federal navy, crediting it with setting up McClellan's new base at Harrison's Landing and concluding that "this shift in bases was a striking demonstration of the flexibility that sea power afforded Northern armies" ("History" 279).

For his land security force, McClellan left nearly half his infantry, with strong artillery support, positioned in two lines directly across the path of the pursuing Confederates. Half the infantry, most of the artillery, and all the wagons began the withdrawal, while two and a half corps were moved to alternate positions, still oriented toward Richmond. These forces were commanded by Generals Sumner and Franklin (Dowdey, *Seven Days* 269).

McClellan felt he was outnumbered, and such a sizable security force indicates his concern for the main body's safety. This concern proved to be well founded when on June 29 Magruder collided with Sumner at Savage's Station. A vicious fight followed, but the Confederate attack was poorly executed. For the time being at least, the Federals thwarted pursuit.

Federal divisions under Generals Smith and Richardson then acted as the rear guard south of White Oak Swamp Bridge while the rest of the security force established positions at Glendale. McClellan and his main body continued to withdraw to the south, but they would not go unmolested. Especially sharp fighting occurred at Glendale on June 30 when Generals Longstreet and Hill struck from the west while Jackson continued his slow approach south. To counter this threat, the Federals had five divisions oriented west and two oriented north. Four other divisions continued to withdraw. This combat power proved sufficient. Of his six divisions, General Lee only got two into the fight, and these could not fix the Federals. McClellan continued moving toward the river and the protection of the gunboats (Dowdey, *Seven Days* 286).

McClellan's next stop was the key terrain at Malvern Hill. The position was naturally defendable, with steep slopes flanked by creeks and within easy range of the gunboats. In fact, here the Federals would benefit from "a perfect shower of shells of tremendous proportion and hideous sound hurled from the heavy guns of the Federal fleet on the James River" (Freeman, *Lee's Lieutenants* 1:583). McClellan began fortifying this position in preparation for one last Confederate attack. On July 1, this attack failed, allowing McClellan to withdraw unimpeded to Harrison's Landing.

Thus, McClellan's security force effectively covered the main body's withdrawal. It was task organized with enough combat power to complete its mission, and McClellan moved his forces to alternate positions to deny pursuit. The security force fought repeated delaying actions at Savage's Station, Glendale, and Malvern Hill, each time withdrawing before becoming decisively engaged. Naval forces provided effective covering fires during these delaying operations. The Confederates had several opportunities to envelop the withdrawing main body but could never succeed. The Federal security force accomplished its first mission, covering the main body.

The other mission of the security force is to deceive the enemy into thinking that no withdrawal is taking place. To facilitate this, current doctrine recommends that displacement during a withdrawal be done by echeloning of forces (FM 3-0 8-6). The Federals again succeeded, with McClellan's withdrawal executed, if not necessarily planned, according to this concept. The move began around noon on June 28 with Keyes's corps moving across White Oak Swamp toward the James River. Porter's corps, which had reorganized after heavy fighting the day before, followed Keyes. Slocum's division would later move to White Oak Swamp to provide additional protection for the wagon trains. Sumner's and Heintzelman's corps occupied fortifications to the front, with Smith's division of Franklin's corps. These troops then moved from their forward positions to the interior lines under the cover of darkness (Dowdey, *Seven Days* 255–56). The secrecy of the movement was further enhanced by a heavy rainfall, which deadened the sound of the withdrawal (Long 176).

This entire procedure went undetected by Confederate pickets, who were, in many places, less than half a mile from Federal lines (Long 175). The Federal withdrawal was not discovered until June 29, when two of Longstreet's engineers found the vacated positions on a reconnaissance (Longstreet 130). Brigadier General Richard Taylor summed up the Confederates' frustration by lamenting, "People find a small cable in the middle of the ocean, a thousand fathoms below the surface [but] for two days we lost McClellan's great army in a few miles of woodland and never had any definite knowledge of its movements" (Hattaway and Jones 195).

Finally, current doctrine cites several considerations and techniques for all retrograde operations, including destroying bridges that might be used by a pursuing enemy, withdrawing nonessential elements prior to withdrawing the main body, and conducting reconnaissance. The Federal withdrawal to Harrison's Landing is full of examples of all these techniques.

McClellan needed good roads to support the withdrawal of his wagons. As early as June 15, he dispatched his topographical engineers to reconnoiter the ground between Richmond and the York River Railroad and White Oak Swamp. He also had his cavalry reconnoiter the known roads leading to the James (Hassler, *McClellan* 135).

The key reconnaissance discovery, however, occurred on June 29 when Keyes found a woods road not listed on any maps that ran parallel to the vital Quaker Road and east of it, toward Malvern Hill (Hassler, *McClellan* 155). Prior to the discovery of this road, the withdrawal had become hopelessly bottlenecked near Glendale. Keyes's discovery allowed many of the wagons to be diverted to the woods road, thus leaving the main road open for troop traffic (Sears, *McClellan* 217; Dowdey, *Seven Days* 283).

While McClellan's men were finding new roads for themselves, they were also closing old ones to the Confederates. The most crucial of these involved the bridge at White Oak Swamp. On June 30, Franklin and his men destroyed the bridge and blocked the road with an abatis. Jackson was already late in crossing the swamp, and this obstacle temporarily slowed him even more. Huger was also slowed by abatis on his way to Glendale. Lee specifically cites McClellan's creation of obstacles as a reason for the Confederate failure to cut off the escape, writing, "Prominent among these [reasons] is the want of correct and timely information. This fact, attributable chiefly to the character of the country, enabled General McClellan skillfully to conceal his retreat and *to add much to the obstructions with which nature had beset the way of our pursuing columns*" (Freeman, *R. E. Lee* 2:232; emphasis added). It is a military truism that the terrain is neutral until someone decides to use it. The terrain was inhospitable and unknown to both Lee and McClellan during the withdrawal. McClellan used this fact to his advantage.

However, even a withdrawal as good as McClellan's falls short of perfection. Perhaps the most critical aspect of the operation was McClellan's failure to evacuate large stores of supplies. While he did move four thousand wagons and a herd of cattle early in the withdrawal, McClellan left behind numerous stockpiles of ammunition, medicine, and baggage. "All articles not indispensable to the safety or the maintenance of the troops" were left behind, including twenty-five hundred nonambulatory wounded who were left in a field hospital in Confederate hands (Dowdey, *Seven Days* 254; Sears, *McClellan* 216).

McClellan took care to destroy everything of military value, a luxury he could not have afforded had the North not had access to such abundant resources (Hattaway and Jones 197). When he gave the order to withdraw, McClellan had twenty-five thousand tons of "essential" supplies to move by land, twenty-five thousand horses and mules, five thousand wagons, and twenty-five hundred cattle (Hassler, *McClellan* 150). In addition to learning a lesson about withdrawing nonessential items early, the Army of the Potomac learned to travel lighter in the future.

Bruce Catton concludes that McClellan and his army conducted their withdrawal from White House to Harrison's Landing with "consummate skill" (*Army of the Potomac* 136). Even McClellan's adversary, Longstreet, praised "McClellan's masterly retreat" (151). These assessments are true when judged not just by Civil War standards but also by current military doctrine. The maneuver has stood the test of time as a classic example of a withdrawal under enemy pressure.

AFTERMATH

With his army safe under the protection of Goldsborough's gunboats, McClellan refused to admit that he had barely escaped destruction. Instead, in a letter to his wife, he boasted, "We have accomplished one of the grandest operations of Military History" (Sears, *McClellan* 225). He dispatched his father-in-law and chief of staff, Randolph Marcy, to Washington to answer *questions* about the campaign and the army's needs. McClellan had given Marcy a letter requesting "rather much over than much less than 100,000 men" (Sears, *McClellan* 223).

After Marcy's briefing, Lincoln decided to come inspect the army. On July 8, he met with McClellan at Harrison's Landing but remained largely silent while McClellan talked. McClellan took the opportunity to present Lincoln with a letter outlining a restrictive war fought against the Confederate army and its government. No effort was to made to subjugate the secessionists or confiscate their property, including slaves. McClellan opined, "A declaration of radical views, especially upon slavery, will rapidly disintegrate our present armies" (Hassler, *McClellan* 177).

McClellan understood "that military matters could not be considered in a vacuum, and that political matters had to be considered also" (Hassler, *McClellan* 178). However, his proposal reflected a fatal inability to discern the changing nature of the war. At the same time that Lincoln was considering expanding the parameters of the war, McClellan was advocating restricting its scope. McClellan did not understand that it was his job to synchronize his military strategy with the political strategy, rather than vice versa. Moreover, McClellan's timing could not have been worse. In the wake of what all around him perceived as an ignoble defeat, McClellan was in no position to be offering such bold advice (Glatthaar 80–82). Consequently, on August 3, Henry Halleck instructed McClellan to return the army to a position south of Washington. Ultimately, most of the Army of the Potomac would fall under the command of John Pope.

T. Harry Williams argues that this was a mistake: "Lincoln—and Halleck and Stanton—would have done better to have left the army where it was. It was only twenty-five miles from Richmond and on a supply line that could always be kept open. It was closer to Richmond than it would be until 1864. Seldom if ever in military history has an army that near to an enemy capital returned without the enemy firing a shot at it. Lincoln would have made a wiser decision if he had kept the army on the James and removed McClellan as its commander" (*Lincoln* 145).

Instead, the Army of the Potomac abandoned the Peninsula, and, as soon as Lee was certain of this development, he went after Pope (Wheeler 352; Williams, *History* 280). Lee again would show his ability to operate within his enemy's decision cycle. McClellan's withdrawal from the Peninsula was no faster than his advance to it. Lee ordered the Army of Northern Virginia to move on the day of the Federal withdrawal, and before the first divisions of the Army of the Potomac had landed at Aquia Creek, Confederate forces had raced north to meet Pope, just south of Manassas, and had him surrounded (Murfin 60).

In the upcoming battle, Lee would apply lessons he had learned during the Seven Days. Key among these Lee cites as "not attacking [the Federals] in their strong and chosen positions. They ought always to be turned." He further explained to Jackson that it was "to save you the abundance of hard fighting that I ventured to suggest for your consideration not to attack the enemy's strong points, but to turn his position. . . . I would rather you have easy fighting and heavy victories" (Jones 71). At Second Manassas, Lee and Jackson would make John Pope the first victim of this wisdom.

Ironically, McClellan used almost the same language as Lee, declaring the "intention of gaining success by maneuvering rather than fighting" (Murfin 84). McClellan's Peninsula Campaign was a well-planned maneuver based on the turning movement Lee also favored. The problem was that McClellan never understood that maneuver can never completely replace fighting and that in the end, war means fighting.

APPENDIX A

CHRONOLOGY OF EVENTS

March 9, 1862	*Monitor* and *Virginia* fight the battle of Hampton Roads.
March 17, 1862	Army of the Potomac commences embarkation from Alexandria, Virginia to Fort Monroe.
March 27, 1862	General Joseph E. Johnston ordered to reinforce Major General John B. Magruder's Army of the Peninsula.
April 1–2, 1862	Army of the Potomac's headquarters transferred to Fort Monroe.
April 4, 1862	Brigadier General Irwin McDowell's First Corps separated from the Army of the Potomac and merged into Department of the Rappahannock. Third Corps and Fourth Corps advance up the Peninsula toward Yorktown. Skirmish at Howard's Mill and Young's Mill.
April 5, 1862	Brigadier General Erasmus Keyes's Fourth Corps halts at Lee's Mill along the Warwick River.
April 5–May 3, 1862	Siege of the Warwick-Yorktown line.
April 12, 1862	Johnston receives command of the Departments of Norfolk and the Peninsula.
April 16, 1862	Battle of Dam no. 1.
April 22, 1862	Brigadier General William Franklin's division arrives and is held in reserve near Ship's Point.
May 3, 1862	Confederates evacuate the Warwick-Yorktown line.
May 4, 1862	Federal cavalry skirmish with Confederates at Lee's Farm and near Williamsburg.
May 5, 1862	Battle of Williamsburg.
May 6, 1862	Federal army occupies Williamsburg.
May 7, 1862	Battle of Eltham's Landing.
May 8, 1862	Federal fleet shells Sewell's Point.
May 8–June 9, 1862	Jackson's Valley Campaign.
May 9, 1862	Confederates evacuate Norfolk and torch Gosport Navy Yard.
May 10, 1862	Federal forces occupy Norfolk and Portsmouth.
May 11, 1862	CSS *Virginia* scuttled by crew off Craney Island.
May 15, 1862	Battle of Drewry's Bluff.

May 18, 1862	Brigadier General Fitz John Porter given command of the Fifth Corps, and Brigadier General Franklin assumes command of the Sixth Corps.
May 31–June 1, 1862	Battle of Seven Pines.
June 1, 1862	General Robert E. Lee replaces Johnston as commander of the Army of Northern Virginia.
June 6, 1862	Major General John Dix replaces Major General John Wool as Federal commander at Fort Monroe, and Dix's ten thousand men are placed under McClellan's command. Wool is reassigned as commander of the Middle Department.
June 12–16, 1862	Brigadier General Jeb Stuart's cavalry rides around the Army of the Potomac and discovers the unprotected Federal right flank.
June 17, 1862	Major General Stonewall Jackson's troops leave the Shenandoah Valley for the Peninsula.
June 25–July 1, 1862	Seven Days battles.
June 25, 1862	Battle of Oak Grove.
June 26, 1862	Battle of Beaver Dam Creek.
June 27, 1862	Battle of Gaines' Mill.
June 29, 1862	Battle of Savage's Station.
June 30, 1862	Battle of Glendale.
July 1, 1862	Battle of Malvern Hill.
July 30, 1862	Army of the Potomac evacuates the sick and wounded from Harrison's Landing.
August 6, 1862	Skirmish at Malvern Hill.
August 14–15, 1862	Third and Fourth Corps move from Harrison's Landing to Aquia Creek.
August 14–19, 1862	Cavalry operations cover the rear of the Army of the Potomac from Harrison's Landing to Williamsburg.
August 20, 1862	Fifth Corps departs from Newport News.
August 21, 1862	Third Corps departs from Yorktown.
August 23, 1862	Sixth Corps departs from Fort Monroe.
August 26, 1862	Second Corps departs from Fort Monroe.

APPENDIX B

TOURING THE BATTLEFIELDS TODAY

The Peninsula Campaign sites encompass three geographical areas: Hampton Roads, Yorktown, and Richmond. The Hampton Roads area sites include Fort Monroe, the *Monitor-Merrimack* Overlook, and the Mariners' Museum. The Yorktown area sites include Lee Hall Mansion, Lee's Mill, Skiffes Creek, Dam no. 1, Endview, Yorktown, and Gloucester Point. The Richmond area sites include the Richmond National Park headquartered at Tredegar Iron Works and the associated sites at Drewry's Bluff, Chickahominy Bluffs, Beaver Dam Creek, Gaines' Mill, Malvern Hill, and Harrison's Landing at Berkeley Plantation.

HAMPTON ROADS AREA

FORT MONROE

Begin the Peninsula Campaign tour at the Federal embarkation site. Today, Fort Monroe remains an active army post. It is the headquarters for the U.S. Army Training and Doctrine Command as well as the home of several other activities. Access to the post is restricted, and visitors must obtain a pass at the security checkpoint. Visitors can walk the nineteenth-century ramparts and tour the Casemate Museum, which provides a comprehensive history of the fort. For more information, write the Casemate Museum, P.O. Box 51341, Fort Monroe, VA 23651 or call 757-788-3391.

MONITOR-MERRIMACK OVERLOOK

Located at Sixteenth Street and Walnut Avenue in Newport News, the site overlooks the March 9, 1862, engagement between the Federal and Confederate ironclads. This stop provides a good panoramic view of Hampton Roads, Sewell's Point, and Craney Island and is marked with a Civil War Trails interpretive sign.

165

THE MARINERS' MUSEUM

The Mariners' Museum, one of the nation's premier nautical museums, houses many important artifacts from the USS *Monitor* and the CSS *Virginia*. The National Oceanic and Atmospheric Administration staff located on-site is preserving recently recovered artifacts from the *Monitor* and is involved in the salvage operation. For more information, write the Mariners' Museum, 100 Museum Drive, Newport News, VA 23606 or call 1-800-581-7245.

YORKTOWN AREA
LEE HALL MANSION

The mansion is located west of the Hampton Roads sites along I-64 near the junction of US 60 and VA 238. It is located near several Peninsula Campaign sites and served as the headquarters for Major General John Magruder and General Joseph Johnston during the siege of the Warwick-Yorktown line. Completed by Richard D. Lee in 1859, Lee Hall has Italianate, Greek-Revival, and Gothic architectural features. Now operated as a historic house museum by the City of Newport News, Lee Hall exhibits rare artifacts in the 1862 Peninsula Campaign Gallery and offers guided tours of the two upper floors. The coauthor of this volume, J. Michael Moore, serves as the registrar. A Civil War Trails interpretative sign is located near the Confederate redoubt on the lawn. For more information, write Lee Hall Mansion, 163 Yorktown Road, Newport News, VA 23603 or call 757-888-3371.

LEE'S MILL

Located at 180 Rivers Ridge Circle in Newport News, Lee's Mill is near the site of Richard D. Lee's grist mill. The Confederates defended this portion of the Warwick-Yorktown line on April 5, 1862, and halted the Army of the Potomac's advance up the Great Warwick Road. The formidable earthworks and fierce Confederate resistance influenced Major General George McClellan's decision to besiege the Warwick-Yorktown line. The site is marked by a Civil War Trails interpretative sign, and interpretative trails follow the Confederate fortifications.

SKIFFES CREEK

The site of several redoubts constructed by Major General John Magruder for protection against a possible Federal flank attack. There is a Civil War Trails interpretative sign near the earthwork.

DAM NO. 1

Located within Newport News Park at 13560 Jefferson Avenue, Dam no. 1 was the only major engagement during the siege of the Warwick-Yorktown line. On April 16, 1862, the Third Vermont Infantry crossed the Warwick River and broke through the Confederate lines. The battle proved a lost opportunity to force a Confederate withdrawal toward Richmond. Extensive park trails follow the Confederate and Federal earthworks. J. Michael Moore, "That Dam Failure: The Battles of Lee's Mill and Dam no. 1" (*North and South* 5 [July 2002]: 62-71) is worth reading before visiting.

ENDVIEW PLANTATION

Endview Plantation, constructed circa 1769, served as a muster ground for the local volunteers in the American Revolution, War of 1812, and the Civil War. Both the Confederate and Federal armies established field hospitals on this site. The site is owned and operated by the City of Newport News. Visitors can tour the exhibit gallery in the English basement and take a guided tour of the upstairs rooms. Walking trails and interpretative signs are located along the grounds. For more information, write Endview Plantation at 362 Yorktown Road, Newport News, VA 23603 or call 757-887-1862.

COLONIAL NATIONAL HISTORICAL PARK

While the park concentrates on the 1781 siege, Yorktown's Civil War history is still apparent in the landscape. The Confederate army redug and expanded the British fortifications surrounding Yorktown and the bluffs commanding the York River. In addition, Major General John Magruder was headquartered at the Customs House, and the Nelson House served as a hospital. For more information, write Colonial National Historical Park, P.O. Box 210, Yorktown, VA 23690 or call 757-898-3400.

GLOUCESTER POINT

Across the York River from Yorktown, Gloucester Point provides visitors with a good view of the commanding bluffs on either side of the river. Walking paths follow the remaining Confederate earthworks, and the site is marked with a Civil War Trails interpretative sign on the Gloucester side of the Route 17 bridge just north of the toll plaza.

RICHMOND AREA
RICHMOND NATIONAL BATTLEFIELD PARK

The Park Service maintains units at Chickahominy Bluffs, Beaver Dam Creek, Gaines' Mill, Glendale, Malvern Hill, and Drewry's Bluff. A complete tour of the park, which includes the 1862 and 1864 battles, involves a drive along an eighty-mile route. Other key sites not part of the park (Seven Pines, Oak Grove, Savage's Station, and White Oak Swamp) are marked by state historical markers. For more information, write Richmond National Battlefield Park, 3215 E. Broad Street, Richmond, VA 23223 or call 804-226-1981.

TREDEGAR IRON WORKS

Established in 1837, Tredegar was the largest iron foundry south of the Potomac River in 1861. It rolled the iron plate for the CSS *Virginia* and the three ironclads of the James River Squadron. In addition, Tredegar cast eleven hundred cannon for the Confederacy and manufactured thousands of artillery projectiles. Tredegar is now the headquarters for the Richmond National Battlefield Park, and the new visitor center includes information on the 1862 and 1864 campaigns, historic Richmond, and area battlefields. It is located at 470 Tredegar Street in Richmond.

DREWRY'S BLUFF

Downriver from Richmond, the bluff still commands an excellent view of the James River. On May 15, 1862, Confederate batteries halted the advance of the USS *Monitor* and several gunboats. Moreover, Drewry's Bluff was an anchorage for the James River Squadron, the site of the Confederate Naval Academy, and a major encampment for the Confederate States Marine Corps. A walking trail leads visitors through the fortifications and the site of the Naval Academy.

CHICKAHOMINY BLUFFS

These bluffs were a strategic location within the Richmond defenses. The high ground overlooks Mechanicsville and the Chickahominy River Valley. From this location, General Robert E. Lee observed the start of the

Seven Days battles. The Park Service maintains walking trails and provides a Civil War Trails interpretative sign.

BEAVER DAM CREEK

The overgrown banks no longer provide an appreciation of the clear fields of fire that allowed Federal artillery and infantry to inflict 1,484 Confederate casualties, but visitors can still appreciate the strength of the Federal position by walking the trails on both sides of the creek. The site is marked with a Civil War Trails interpretative sign.

GAINES' MILL

Gaines' Mill was General Robert E. Lee's only tactical victory during the Seven Days battles. Visitors can follow the expanded walking trail to the point where Texas and Georgia troops broke the Federal lines around nightfall and can read the numerous interpretative signs along the path.

MALVERN HILL

On July 1, 1862, Confederate infantry advanced across open ground against an impregnable Federal position on a 150-foot slope flanked by swamps and river bottoms. More than a hundred Federal cannon massed along the slope mowed down the Confederates and inflicted half of the 5,355 casualties. The Park Service recently acquired further acreage and added more interpretative trails. Most importantly, the Park Service restored the landscape's 1862 appearance, which affords the visitor with the same view as the Federal gunners had of the Confederate columns.

BERKELEY PLANTATION (HARRISON'S LANDING)

This plantation is located on the James River, thirty-five miles east of Richmond and eighteen miles west of Williamsburg on Route 5. During the Peninsula Campaign, it served as the supply base for the Army of the Potomac. After the Seven Days battles, Federal troops, protected by gunboats, rested on the grounds of this 1726 structure. In addition, Daniel Butterfield composed "Taps" while stationed here. On July 8, 1862, President Abraham Lincoln visited with the troops and conferred with Major General George McClellan. Guided tours of Berkeley Plantation are offered throughout the year, and visitors can walk around the extensive gardens. For more information, write Berkeley Plantation, 12602 Harrison Landing Road, Charles City, VA 23030 or call 804-829-6018 or 888-466-6018.

APPENDIX C

TIPS FOR CONDUCTING A STAFF RIDE

The Staff Ride by William Robertson is perhaps the most comprehensive guide to the conduct of staff rides available. In it, Robertson writes, "A staff ride consists of systematic preliminary study of a selected campaign, an extensive visit to the actual sites associated with that campaign and an opportunity to integrate the lessons derived from each. It envisions maximum student involvement before arrival at the site to guarantee thought, analysis, and discussion. A staff ride thus links a historical event, systematic preliminary study, and actual terrain to produce battle analysis in three dimensions" (5).

Robertson lists several objectives of a staff ride:

a. To expose students to the dynamics of battle, especially those factors that interact to produce victory and defeat.
b. To expose students to the "face of battle," the timeless human dimensions of warfare.
c. To provide case studies in the application of the principles of war.
d. To provide case studies in the operational art.
e. To provide case studies in combined arms operations or in the operation of a single arm or branch.
f. To provide case studies in the relationship between technology and doctrine.
g. To provide case studies in leadership at any level desired.
h. To provide case studies in unit cohesion.
i. To provide case studies in how logistical considerations affect operations.
j. To show the effects of terrain on plans and their implementation.
k. To provide an analytical framework for the systematic study of campaigns and battles.
l. To encourage officers to study their profession through the use of military history.
m. To kindle or reinforce an interest in the (military) heritage (Robertson 5–6).

The Peninsula Campaign lends itself very well to providing a venue for accomplishing the staff ride objectives described by Robertson, and this volume has been designed to facilitate this goal. For example, the discussion of Jackson's exhaustion after the Valley Campaign and its impact on his performance during the Peninsula Campaign can be used as a case study of the human dimension of warfare. Similarly, the principles of war are used to analyze McClellan's conduct of the campaign and the facets of operational art are used to analyze Lee's actions.

171

The Peninsula Campaign provides an excellent case study of the difficulties in achieving unity of effort during joint operations, and in this regard the campaign captures the joint aspect of Robertson's army-focused objective of studying combined arms operations. The use of such innovations as ironclads, mines, and balloons during the Peninsula Campaign provides an opportunity to study the relationship between technology and doctrine.

The biographical portion of this volume is designed to help with the study of leadership. McClellan's use of the James and York Rivers as his lines of communication and his change of base show how logistical considerations affect operations. The various discussions of terrain, especially as to how it hindered McClellan's initial offensive and strengthened Porter's defense at Mechanicsville and Malvern Hill, show the effects of terrain on plans and their implementation.

Robertson identifies three distinct phases of the staff ride: preliminary study, field study, and integration (5). By providing the background information necessary to achieve Robertson's objectives, this volume is designed to be a useful tool during the preliminary study phase of a staff ride.

This volume also provides a section on "Touring the Battlefield Today" that is designed to assist with the field study phase. It is also recommended that this phase of the staff ride include an early visit to Lee Hall Mansion in Newport News, Virginia, where coauthor J. Michael Moore works as the on-site historian. With prior coordination, Mr. Moore would be delighted to assist with a staff ride.

The final phase of the staff ride is integration. This is an after-action review of the entire staff ride and an opportunity to derive lessons leaned. Robertson writes that the student must achieve what Clausewitz "defined as *critical analysis:* determine the facts, establish cause and effect, and analyze results. In simpler terms, the soldier must find out what happened, establish why and how events occurred as they did, and decide what these cause and effect relationships mean. It is the immediacy of this last element—the answer to the question, 'So what?'—that makes this approach to battle analysis a particularly military endeavor. The effect of such analysis is synergistic in fostering not just lessons but a deeper understanding of the realities of war" (4). This should be the goal of the integration phase, and chapter 5 of this volume will assist in this effort.

Former Chief of Staff of the Army General Carl Vuono writes that "nowhere is [the] close connection between history and training more apparent than in the staff ride. . . . It is the one training technique that uses military history with the actual battlefield to bring together the realities of war" (1). The proximity of the Peninsula Campaign battlefields to the Joint Forces Staff College, Fort Eustis, Fort Monroe, Naval Amphibious Base Little Creek, and other military activities in the Virginia area make the lessons of the staff ride easy to integrate into joint and service military instruction. The Peninsula Campaign affords an excellent vehicle for such study, and this volume is designed to be a tool used to this end.

BIBLIOGRAPHY

Alexander, Bevin. *Lost Victories: The Military Genius of Stonewall Jackson*. New York: Holt, 1992.

Alexander, Edward Porter. *Fighting for the Confederacy*. Chapel Hill: University of North Carolina Press, 1989.

Anderson, Bern. *By Sea and by River: A Naval History of the Civil War*. Westport, Conn.: Greenwood, 1962.

Bailey, Ronald. *Forward to Richmond: McClellan's Peninsular Campaign*. Alexandria, Va.: Time-Life, 1983.

Boatner, Mark. *The Civil War Dictionary*. New York: McKay, 1959.

Catton, Bruce. *Army of the Potomac: Mr. Lincoln's Army*. Garden City, N.Y.: Doubleday, 1962.

————. *The Civil War*. New York: American Heritage, 1960.

————. *Terrible Swift Sword*. Garden City, N.Y.: Doubleday, 1963.

————. *This Hallowed Ground*. Garden City, N.Y.: Doubleday, 1956.

Clausewitz, Carl von. *On War*. Ed. Michael Howard and Peter Paret. Princeton: Princeton University Press, 1976.

Cullen, Joseph. *The Peninsula Campaign 1862: McClellan and Lee Struggle for Richmond*. Harrisburg, Pa.: Stackpole, 1973.

Davis, Burke. *The Civil War: Strange and Fascinating Facts*. New York: Fairfax, 1982.

————. *Jeb Stuart: The Last Cavalier*. New York: Bonanza, 1957.

Deaderick, Barron. *Strategy in the Civil War*. Harrisburg, Pa.: Military Service Publishing, 1946.

Donovan, Tim, et al. *The American Civil War*. West Point: U.S. Military Academy, 1980.

Dougherty, Kevin. "Drewry's Bluff: A Blocking Position." *Infantry* 80 (January–February 1990): 20–21.

————. "Simple Scraps of Cloth." *Soldiers* 48 (March 1993): 52.

Dowdey, Clifford. *The Land They Fought For*. Garden City, N.Y.: Doubleday, 1955.

————. *The Seven Days*. New York: Fairfax, 1978.

Dowdey, Clifford, and Louis Manarin, ed. *The Wartime Papers of R. E. Lee*. New York: Bramhall House, 1961.

Eckenrode, H. J., and Bryan Conrad. *James Longstreet: Lee's War Horse*. Chapel Hill: University of North Carolina Press, 1986.

Faust, Patricia, ed. *Historical Times Illustrated Encyclopedia of the Civil War*. New York: Harper and Row, 1986.

173

Fleek, Sherman. "Fitz-John Porter: A Scapegoat for Disaster." *Army* 44 (February 1994): 52–60.

FM 3-0. *Operations*. Washington, D.C.: Department of the Army, 2001.

FM 7-8. *Infantry Rifle Platoon and Squad*. Washington, D.C.: Department of the Army, 1992.

FM 22-100. *Army Leadership*. Washington, D.C.: Department of the Army, 1999.

FM 101-5. *Staff Organization and Operations*. Washington, D.C.: Department of the Army, 1997.

FM 101-5-1. *Operational Terms and Graphics*. Washington, D.C.: Department of the Army and the U.S. Marine Corps, 1997.

Foote, Shelby. *The Civil War: A Narrative*. Vol. 1. New York: Random House, 1958.

Freeman, Douglas Southall. *Lee's Lieutenants: A Study in Command*. 3 vols. New York: Scribner's, 1942–44.

———. *R. E. Lee*. 4 vols. New York: Scribner's, 1934.

Fuller, J. F. C. *Generalship: Its Diseases and Their Cure; A Study of the Personal Factor in Command*. Harrisburg, Pa.: Military Service Publishing, 1936.

Glass, Scott. "Battle of Beaver Dam Creek: FM 100-5 Lessons Learned." *Infantry* 84 (November–December 1994): 10–14.

Glatthaar, Joseph. *Partners in Command: The Relationship between Leaders in the Civil War*. New York: Free Press, 1994.

Guttman, Jon. "Rebel Stand at Drewry's Bluff." *America's Civil War*, November 1997, 1–7.

Goss, Warren. "Yorktown and Williamsburg." In *Battles and Leaders of the Civil War*, 2:189–99. Edison, N.J.: Castle, n.d.

Halsey, Ashley. *Who Fired the First Shot?* New York: Crest, 1983.

Harris, Shawn. "Lion's Tail Touched." *Military History* 15 (June 1998): 42–48.

Hassler, Warren. *Commanders of the Army of the Potomac*. Baton Rouge: Louisiana State University Press, 1962.

———. *General George B. McClellan: Shield of the Union*. Baton Rouge: Louisiana State University Press, 1957.

Hattaway, Herman, and Archer Jones. *How the North Won*. Chicago: University of Illinois Press, 1983.

Joint Publication 3-0. *Doctrine for Joint Operations*. Washington, D.C.: Joint Chiefs of Staff, February 1995.

Joint Publication 3-02. *Joint Doctrine for Amphibious Operations*. Washington, D.C.: Joint Chiefs of Staff, October 1992.

Joint Publication 3-02.1. *Joint Doctrine for Landing Force Operations*. Washington, D.C.: Joint Chiefs of Staff, November 1989.

Joint Publication 3-13. *Joint Doctrine for Information Operations*. Washington, D.C.: Joint Chiefs of Staff, 1998.

Joint Publication 3-58. *Joint Doctrine for Military Deception*. Washington, D.C.: Joint Chiefs of Staff, 1996.

Jomini, Antoine-Henri de. *Summary of the Art of War*. New York: Putnam, 1854.

Jones, Archer. *Civil War Command and Strategy*. New York: Free Press, 1992.

Liddell Hart, Basil Henry. *Strategy*. New York: New American Library, 1974.

Long, A. L. *Memoirs of Robert E. Lee*. Secaucus, N.J.: Blue and Grey, 1983.

Longstreet, James. *From Manassas to Appomattox*. Secaucus, N.J.: Blue and Grey Press, 1984.

McClellan, George. "The Peninsular Campaign." In *Battles and Leaders of the Civil War*, 2:160–88. Edison, N.J.: Castle, n.d.

McWhiney, Grady, and Perry Jamieson. *Attack and Die*. Tuscaloosa: University of Alabama Press, 1982.

Murfin, James. *The Gleam of Bayonets*. Baton Rouge: Louisiana State University Press, 1982.

National Park Service Historical Handbook: Richmond Battlefields. Available at http://www.cr.nps.gov/history/online-books/hh/ee/hh33ehtm. Accessed June 17, 2004.

Nevins, Allan. *The War for the Federal: War becomes Revolution, 1862–1863*. New York: Scribner's, 1960.

Niven, John. *Gideon Welles: Lincoln's Secretary of the Navy*. New York: Oxford University Press, 1973.

O'Neill, Robert. "Cavalry on the Peninsula." *Blue and Gray Magazine* 19 (June 2002): 6–51.

Parish, Peter. *The American Civil War*. New York: Holmes and Meier, 1975.

Pollard, E. A. *The Lost Cause*. New York: Treat, 1867.

Quarstein, John V. *The Battle of the Ironclads*. Charleston, S.C.: Arcadia, 1999.

Quarstein, John V., and Dennis Mroczkowski. *Fort Monroe: Key to the South*. Charleston, S.C.: Arcadia, 2000.

Reed, Rowena. *Combined Operations in the Civil War*. Annapolis, Md.: Naval Institute Press, 1978.

Roberts, Robert. *Encyclopedia of Historic Forts*. New York: Macmillan, 1988.

Robertson, James I. *Stonewall Jackson: The Man, the Soldier, the Legend*. New York: Macmillan, 1997.

Robertson, William. *The Staff Ride*. Washington, D.C.: Center of Military of History, 1987.

Schenck, Martin. *Up Came Hill: The Story of the Light Division and Its Leaders*. Harrisburg, Pa.: Stackpole, 1958.

Sears, Stephen. *Chancellorsville*. Boston: Houghton Mifflin, 1996.

———. *George B. McClellan: The Young Napoleon*. New York: Ticknor and Fields, 1988.

———. *To the Gates of Richmond: The Peninsular Campaign*. New York: Ticknor and Fields, 1992.

Smith, Stuart. *Douglas Southall Freeman on Leadership*. Shippensburg, Pa.: White Mane Publishing, 1993.

Stiles, T. J. *In Their Own Words: Civil War Commanders*. New York: Perigee, 1995.

Stuckey, Scott. "Joint Operations in the Civil War." *Joint Forces Quarterly* 6 (autumn–winter 1994–95): 92–105.

Sun Tzu. *The Art of War*. Trans. Thomas Cleary. Boston: Shambhala, 1988.

Tanner, Robert. *Stonewall in the Valley*. Garden City, N.Y.: Doubleday, 1976.

Thomas, Emory M. *The Bold Dragoon: The Life of J. E. B. Stuart*. New York: Vintage, 1986.

———. *Robert E. Lee*. New York: Norton, 1995.

Trefousser, Hans. *The Radical Republicans: Lincoln's Vanguard for Racial Justice*. New York: Knopf, 1969.

Vandiver, Frank. *The Mighty Stonewall*. College Station: Texas A&M University Press, 1957.

Vuono, Carl. "The Staff Ride: Training for Warfighting." *Army Historian* 12 (October 1998): 1–2.

War Department. *The War of the Rebellion: A Compilation of the Official Records of the Federal and Confederate Armies*. 128 vols. Washington, D.C.: U.S. Government Publishing Office, 1884.

Warner, Ezra. *Generals in Blue*. Baton Rouge: Louisiana State University Press, 1964.

Weigley, Russell. *The American Way of War*. Bloomington: Indiana University Press, 1973.

Wert, Jeffry. *General James Longstreet: The Confederacy's Most Controversial Soldier: A Biography*. New York: Simon and Schuster, 1993.

West, Richard. *Lincoln's Scapegoat General: A Life of Benjamin F. Butler, 1818–1893*. Boston: Houghton Mifflin, 1965.

Wheeler, Richard. *Sword over Richmond: An Eyewitness History of McClellan's Peninsula Campaign*. New York: Fairfax, 1986.

Wiley, Bell Irvin. *The Life of Johnny Reb: The Common Soldier of the Confederacy*. Garden City, N.Y.: Doubleday, 1943.

Williams, T. Harry. *The History of American Wars*. New York: Knopf, 1981.

———. *Lincoln and His Generals*. New York: Vintage, 1952.

Wood, John Taylor. "The First Fight of the Iron-Clads." In *Battles and Leaders of the Civil War*, 1: 692–711. Edison, N.J.: Castle, n.d.

INDEX